Environmental business management

An introduction

A training videocassette entitled *The Green Challenge* has been produced by TV Choice Productions (London, 1995) in collaboration with UNEP and the ILO, based on the ideas contained in this book. Copies are available from TV Choice Productions, 22 Charing Cross Road, London WC2H OHR, United Kingdom.

Management Development Series No. 30

Environmental business management

An introduction

Second (revised) edition

Klaus North

International Labour Office   Geneva

North, K.
Environmental business management: An introduction (second edition)
Geneva, International Labour Office, 1997 (Management Development Series, No. 30)

/Guide/, /Environmental protection/, /Management/, /Management development/, /Industrial enterprise/s. 16.03.1
ISBN 92-2-109516-9
ISSN 0074-6703

ILO Cataloguing in Publication Data

ILO publications can be obtained through major booksellers or ILO local offices in many countries, or direct from ILO Publications, International Labour Office, CH-1211 Geneva 22, Switzerland. Catalogues or lists of new publications are available free of charge from the above address.

Contents

Tables

Figures

Introduction

Why are enterprises increasingly trying to become Green and clean? For one company the reason may be a major industrial accident, another may have experienced pressure from Green consumers and a third company may have discovered new markets. The loss of market shares to Green competitors is a good incentive to change. The insight that retrofitting is a more expensive way to fulfil government regulations than considering the environment at the investment decision stage is another starting-point. Even managers' families often play an important role in convincing them that their business is polluting the environment.

While business threatens the environment and our survival by depleting resources and polluting the air, water and land, the importance of "Green and clean" can, in turn, be perceived as a threat by many enterprises. Yet managers are beginning to consider the environment as an additional parameter on which to base their decisions. For the purpose of this book we have defined the environment as the sum of physical resources that sustain life and are the basis for satisfying human needs. Environmental business management is taken to mean the integration of environmental protection into all managerial functions with the aim of reaching an optimum between the economic and ecological performance of a company.

In this respect environmental business management is a key element in achieving sustainable development of industrialized and developing countries.

The attraction of polluter-intensive industries from the North formed, and continues to form, part of the industrial development strategy for a number of developing and newly industrialized countries. It is now becoming clear, however, that this development strategy is negatively affecting not only the country's national heritage, the health of workers and population, but also the long-term competitive advantages of a country.

In contrast, industries which have to comply with strict environmental regulations have developed the know-how in "Green and clean" products and processes, resulting in competitive advantages in national and international markets. As the pressure for stricter environmental regulations on a global level mounts, those countries and industries which were among the first to implement environmental protection measures now have considerable growth opportunities.

This leads to an overall positive effect on productive employment and the quality of working and living conditions. Employers, managers, workers and their families all share a concern for the environment. Environmental business management can only then be meaningful if it is based on an open dialogue between all parties.

During its 77th Session in 1990 the International Labour Conference called upon government, and employers' and workers' organizations to develop concerted strategies to provide adequate education and training to all parties involved in environmental protection. Furthermore, it was requested that the ILO pay increased attention to environmental education and training, by integrating environmental considerations more effectively into training activities related to ILO programmes in all areas. The present book is a contribution to putting the resolution of the International Labour Conference into practice.

This book aims at helping business managers turn environmental threats into growth opportunities for their businesses and also into opportunities for the sustainable development of our planet. These two aspects form an integral part of environmental business management.

In some industrialized countries major polluter companies and consumer product manufacturers had already had to adjust to Green pressure and the new environmental standards in the 1970s. Their managers learned mainly by reacting to threats; for instance, blockages of their company entrances by Green activists, withdrawn authorizations to extend production, and Not In My Backyard (NIMBY) protests at possible new production sites. In the 1970s it was not rare for some industrial sectors to spend roughly 20 per cent of their capital investments in cleaning up their production facilities. For many of those companies the early investments have now turned into profits, and have led to a proactive environmental management strategy.

While reading this book, managers of such companies will perhaps feel satisfied that they have been doing the right thing. They may find additional information on how to integrate environmental considerations into all functional areas of their enterprise.

A second category of managers who could benefit from this book are those who have just been hit by the "environmental wave" and want to know how to react and how to establish proactive policies.

Those managers who are already convinced that the protection of our environment forms part of a manager's responsibilities to society will find examples in this book of how to translate this responsibility into concrete actions.

This book may serve to raise the environmental awareness of two more categories of managers: those who have not yet been confronted with the environmental discussion and those who are more concerned with quarterly profits than with the long-term survival of a company.

The present book may be useful also for officials of environmental protection agencies who wish to learn more about the possibilities of translating environmental regulations into cost-effective measures within industry, for labour inspectors who have been given additional responsibilities for environmental protection, and for course planners at universities and business schools who want to integrate environmental aspects into their business management courses.

Workers' representatives and trade union officials may also be interested in this book as a basis for negotiating company and collective agreements which include environmental aspects.

What can the reader expect to find here?

Chapter 1 outlines the environmental challenge by explaining how companies will be affected by the necessity to protect the environment, what sustainable development means for business, how environmental standards regulate business activities, and how business management can cope with the new challenge. Chapter 1 can also be seen as an executive summary of the book.

Chapter 2 explains how environmental considerations can be integrated into the main management functions of an enterprise.

Chapter 3 consists of two parts: the first part provides management tools to "make your business lean, Green and clean", such as developing an action plan or conducting environmental impact assessments and audits. The second part develops specific issues such as dealing with wastes and pollution, energy saving and preparation for emergencies.

Environmental business management is not limited to isolated actions of individual enterprises. There are many tasks where managers can benefit from outside support: contribution to the establishment of cost-effective environmental standards, transfer of clean technology, countering the polluter image of a business sector, negotiating agreements with trade unions and entering in a constructive dialogue with environmentalist groups. These topics are dealt with in *Chapter 4*.

Useful addresses of institutions are provided in Appendix 3.

For those managers who feel that they need further reading on the subject there is an extensive *bibliography*. *A glossary* explains the main environmental terms and issues such as greenhouse effect, ozone depletion, acidification, etc.

The completion of this book would not have been possible without the assistance of a great number of people. I wish to thank in particular my colleagues Koe Doeleman, Harry Evan, Larry Kohler and Derek Miles who provided material, valuable ideas and critical comments. Michael Royston and Colin Guthrie prepared a draft paper on environmental management, passages of which have been used in some sections, and their ideas have influenced some other sections. Many organizations and companies provided information and insights into their business practices. I would also like to extend my thanks to Beate Stoffers and Astrid Scholz who assisted me in literature research, Lydia Badia and Julia Carrera who typed and prepared the manuscript, and Lilian Neil who edited it.

Finally, the ILO would be most grateful to readers for any comments and suggestions.

1

The environmental challenge

This first chapter tackles the following key issues:
- [ ] how enterprises are affected by the importance of protecting the environment;
- [ ] what the scenario of sustainable development means for business;
- [ ] how environmental standards regulate business activities; and
- [ ] how business management can cope with the environmental challenge.

1.1 The environment: Business opportunities and threats

The key question for a manager, "how to make money and protect the environment at the same time?", can be answered in a variety of ways according to the setting. It not only applies to "Greening and cleaning" existing businesses, but also requires entrepreneurial creativity to turn environmental constraints into new business opportunities.

Recycling has become a multibillion dollar business with high growth rates (Nulty, 1990; Sherman, 1989). Shares for waste management companies are performing well in stock markets. Fund managers are increasingly considering environmental factors in their investment decisions. Specialized "Green" investment funds become new businesses. Providers of clean process technology are booming. A new species, "the environmental consultant", has emerged. Advertising agencies, auditors and law firms have entered the new business (Elkington and Burke, 1989). The environment seems to have become the leading marketing argument of the 1990s. Managers grasp the environment issue as a potential competitive advantage in saturated markets. The prospects of additional new products such as phosphate-free detergents or "ozone friendly" chemicals increase with consumer awareness and national or international regulations.

Some managers may, upon reading this, have reservations and feel that environmental discussions and forthcoming regulations could pose a threat to their company rather than be an opportunity for growth. In some cases they could be right.

For example, by the mid-1980s in Izmir, on the Aegean coast of Turkey, the quantity of contaminated water dumped by local industries into the sea had become unacceptable. By the end of the 1980s the local government,

---

Pollution prevention pays

☐ Since 1975 when the American company 3M pioneered its 3P programme "Pollution Prevention Pays" in the United States, more than 2,500 projects have been carried out leading, according to statements of the company, to savings of over US$500 million.

☐ As a result of its productivity movement the Indian BHEL corporation saved energy worth approximately 26 million rupees within two years.

☐ The investment to recover foundry dust and reuse it to make new moulds proved to be highly successful at Baxi partnership in the United Kingdom. Not only was the investment paid back in a mere three months, but Baxi partnership was also awarded the 1989 prize for good environmental management by the Industry and Environment Office of the United Nations Environment Programme and the Commission of the European Communities.

---

with the help of the World Bank, had begun the construction of a water purification plant. All businesses discharging contaminated effluents must use this facility. In order to keep the total cost of the project within reasonable limits, it was decided that the plant would only carry out secondary purification. The primary purification would be done by each individual enterprise. While this posed no great threat to firms whose effluents were only mildly contaminated, or to large firms who could easily obtain loans to install their own water-treatment plants, it was a major threat to small, dirty firms such as tanneries. It is likely that many of these will be forced to close.

Other major polluters such as pulp and paper, food processing, iron and steel, aluminium, chemicals, cement manufacturing and electric power generation may face costly retrofitting and equipment changes, which may account for between 5 and 15 per cent of total new plant and capital expenditures, as a study for the United States has shown (Leonard, 1988, p. 90).

A manager would certainly like to know how his or her business will be affected as a consequence of the increasing environmental consciousness of consumers and governments. Will the environmental challenge present an opportunity or a threat to the business?

An estimated answer to this question can be reached by assessing the present state of a company by means of a quick **Environmental Challenge Scan**, as presented in figure 1. You may wish to draw an environmental challenge profile of your company by indicating for each of the criteria whether you are closer to the characteristics of a "Green growth" company or to a threatened company. Even though there are many more criteria to define the success of a company, the criteria which we will now discuss briefly are either a major asset or a heavy burden "in making money and protecting the environment at the same time".

By running through this quick assessment of environmental challenges, a manager will become aware of the main problem areas of environ-

---

Business environment interaction: The case of ozone depletion

Chloro-fluorocarbons (CFCs), halons and other substances damage the ozone layer in the stratosphere. Since the early 1980s, when a vast hole in the ozone layer above the Antarctic was discovered, relatively rapid international action has been taken to ban the substances harmful to the ozone layer. Already in 1987, the Montreal Protocol and related national regulations called for phased reduction of CFC production and a complete phasing-out by the year 2000. As the damage to the ozone layer proves to be more dramatic and other substances have been found that also contribute to ozone depletion, the London Amendment (1990) and the Copenhagen Amendment (1992) foresee more drastic and quicker reductions: CFCs and carbon tetrachloride have to be phased out by 1996, while production of methyl bromide was frozen in 1995 at 1991 levels.

This not only results in obsolete capacities in the chemical industry, but also caused a number of users to develop alternative solutions for the products using CFCs and halons. The many users of aerosol sprays had to look for an alternative propellant. Producers of refrigerators are still searching for an equally performative refrigerant. Enterprises have to design services to scrap refrigerators without releasing CFCs. In the production of synthetic materials (insulating and packaging foams), new blowing agents are required. For the cleaning of metal and electronic parts substitutes have to be found. The same is true for halons used in fire extinguishers. The threat of businesses to the environment backfires as a threat to companies using CFCs and halons. Those companies succeeding in developing alternative solutions will have a competitive advantage: a threat turned into an opportunity, a growth potential for innovative companies.

---

mental management, which will be dealt with in more detail in the remaining chapters of the book. In the following, each of the evaluation criteria will be discussed.

Sector of the economy

The sector to which an enterprise belongs is already an indicator both of the threat it poses to the environment and of the costs for enterprises associated with stricter environmental regulations (Leonard, 1988; OECD, 1991).

Table 1 relates sectors of the economy to their possible threats to the environment. The Brundtland report names the following as major polluters: food processing; iron and steel; non-ferrous metals; automobiles; pulp and paper; chemicals; and electro-power generation (WCED, 1987). Taylor et al.

(1994) identify three distinct categories of sector with regard to their environmental concerns and levels of environmental management: first, dirty, damaging and dangerous such as water, energy and mining; second, wasteful and polluting such as light industry and retailing; and, third, the so-called "silent destroyers", e.g. the professions, government and finance. There is also a recent, growing awareness that agriculture is not necessarily a clean business. Other major sources of pollution are private households and road transport. Technologies and production levels, however, vary widely throughout the world. A sectoral analysis can only be a very rough indicator of "pollution prone" activities. For example, a pulp and paper manufacturer has been able, by using advanced technology, to reduce water and energy consumption to a minimum, recycle waste water in a closed circuit, use less harmful bleaching agents for the paper, and base the production on recycling used paper. Another pulp and paper plant heavily pollutes the local river, imports high amounts of wood, and is a consumer of enormous amounts of energy. The whole area has an unpleasant odour. The latter company contributes to the persistence of a polluter image, threatening the environment and the public image of the sector, while the former has successfully reduced its threats to the environment.

Products

The environmental friendliness of an individual enterprise is determined not only by its manufacturing processes but also by the products it yields.

Products should:

☐ have an extended life span;
☐ be made of renewable materials;
☐ not pollute the environment;
☐ be energy efficient in production and use;
☐ require minimal packaging.

In the case of the car, for instance, it is significant to observe the changes that have been introduced as a consequence of energy costs, together with antipollution and antiwaste regulations: drastic reduction of fuel consumption by improved engine design, reduced weight and wind resistance; reduction of exhaust emissions by catalyser technology; and partial use of recyclable materials. Even if each individual car becomes more "environment friendly", it does not necessarily mean that road transport as a whole is less threatening to the environment. We are observing an increasing number of automobiles worldwide and the distance driven per vehicle is also growing. Growth, even with Greener products, may therefore be counterproductive to environmental protection.

Figure 1. The Environmental Challenge Scan

	The threatened company	Company score					"Green growth" company
Sector of economy	Dirty, damaging, dangerous, wasteful and polluting	1	2	3	4	5	Low pollution
Products	Nonrenewable materials Polluting, high consumption of resources						Renewable and recyclable materials Non-polluting Low energy consumption
Processes	Polluting Hazardous wastes High energy consumption Health hazards to workers						Non-polluting Low waste Low energy consumption Efficient use of resources No health hazard to workers
Environmental consciousness	Consumers not environmentally conscious						Environmentally conscious consumers
Environmental standards	Low standards or non-compliance with strict standards						Compliance with strict environmental standards
Management and staff commitment	No commitment						Committed to environmental protection
Skill level of staff	Low, highly specialized in old technologies						High, good general education
R&D capacity	Low R&D profile						Creative team; short development cycle
Capital	Capital shortage						Environmentally conscious financing institutes

Explanation of scoring:
1 = Company highly threatened by environmental debate.
5 = Environmental debate constitutes growth opportunity.

Table 1. Environmental effects of selected industrial sectors

Sector	Raw material use	Air	Water resources		Solid wastes and soil	Risk of accidents	Others: noise, workers' health and safety, consumer products
			Quantity	Quality			
Textiles	Wool, synthetic fibres, chemicals for treating	Particulates, odours, SO_2, HC	Process water	BOD, suspended solids, salts, sulphates, toxic metals	Sludges from effluent treatment		Noise from machines, inhalations of dust
Leather	Hides, chemicals for treating and tanning		Process water	BOD, suspended solids, sulphates, chromium	Chromium sludges		
Iron and steel	Iron ore, limestone, recycled scrap	Major polluter: SO_2, particulates, NO, HC, CO, hydrogen sulphide, acid mists	Process water	BOD, suspended solids, oil, metals, acids, phenol, sulphides, sulphates, ammonia, cyanides, effluents from wetgas scrubbers	Slag, wastes from finishing operations, sludges from effluent treatment	Risk of explosions and fires	Accidents, exposure to toxic substances and dust, noise
Petro-chemicals refineries	Inorganic chemicals	Major polluter: SO_2, HC, NO_x, CO, particulates, odours	Cooling water	BOD, COD, oil, phenols, chromium, effluent from gas scrubbers	Sludges from effluent treatment, spent catalysts, tars	Risk of explosions and fires	Risk of accidents, noise, visual impact
Chemicals	Inorganic and organic chemicals	Major polluter: organic chemicals (benzene, toluene), odours, CFCs		Organic chemicals, heavy metals, suspended solids, COD, cyanide	Major polluter: sludges from air and water pollution treatment, chemical process wastes	Risk of explosions, fires and spills	Exposure to toxic substances, potentially hazardous products
Non-ferrous metals (e.g. aluminium)	Bauxite	Major local polluter: fluoride, CO, SO_2, particulates		Gas scrubber effluents containing fluorine, solids and hydrocarbons	Sludges from effluent treatment, spent coatings from electrolysis cells (containing carbons and fluorine)		
Micro-electronics	Chemicals (e.g. solvents), acids	Toxic gases		Contaminations of solids and groundwater by toxic chemicals (e.g. chlorinated solvents), accidental spillage of toxic material			Risk of exposure to toxic substances
Bio-technologies				Used for effluent treatment	Used for clean-up of contaminated land		Fears of hazards from the release of micro-organisms into the environment

Source: OECD, 1991, p. 189.

Processes

Processes which have earned the label "environment friendly" come close to the objectives:

☐ zero pollution;

☐ zero waste production;

☐ zero health hazards to the workers;

☐ low energy consumption; and

☐ efficient use of resources.

To assess how near an enterprise is to this goal, an estimated environmental balance, relating inputs and outputs, should be established (as illustrated in figure 2). The identification of all inputs and outputs in quantitative terms is virtually impossible for industrial enterprises. It is sometimes easier to determine, for example, the amount of water being used to produce one ton of paper and the quality of the water leaving the company, or the energy needed to produce one ton of steel and the emissions associated with this process.

It has to be kept in mind, however, that even a company which monitors inputs and outputs, and complies with the established environmental standards, does not automatically become environment friendly. Firstly, the long-term consequences of the released substances for the environment may still be unknown and, secondly, environmental standards may generously allow emissions which are above threshold levels, leading to adverse environmental effects.

Environmental consciousness and standards

While environmentally unconscious consumers may give managers the comfortable feeling of not being threatened by the "Green wave", these enterprises may be trapped by companies which have already adjusted to the new challenge. Even though environmentally conscious consumers may cause a loss of market share in the short run, or force an enterprise to withdraw from plans to extend its production because of citizen protests, in the long run they are the real supporters of Green growth.

It can be observed that countries with strong Green movements also have strict environmental regulations.

When dealing with the argument that countries with the most complicated home markets are most likely to be competitive in the world market, we might argue that those countries with the toughest environmental legislation are the ones most likely to be successful in environment-related business. Countries such as Japan, the United States, Germany and Sweden, for example, all of which have adopted comparatively tough environmental legislation, seem to support this. Porter and van der Linde (1995) argue that

Figure 2. The physical interaction of industrial activities and environment

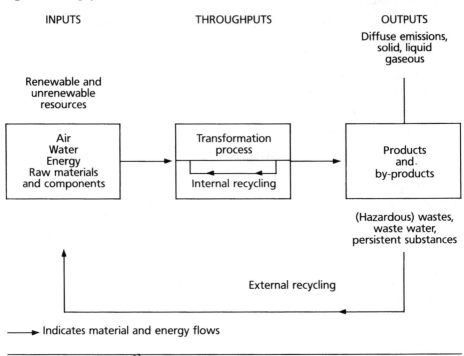

— Indicates material and energy flows

properly determined environmental standards can trigger innovations and lower the total cost of a product or improve its value. They thus advocate a win-win mind-set of environmental management. (See section 1.2 of this book.)

Management and staff commitment

The most important asset in bringing about change is management and staff commitment. In our Environmental Challenge Scan those companies which have already undertaken measures to sensitize managers and staff to the importance of environmental protection score high points. Many of these companies have attempted in their suggestion schemes or quality circles to render products and processes less harmful to the environment. These are invariably companies which have a genuine concern for the working environment.

As the environment is becoming a fashionable business topic, we find many companies paying lip service to environmental protection without taking meaningful actions to prove that they are really committed.

The chief executive officer of an automobile manufacturer proposing to build more four-lane highways in Europe to reduce traffic congestion, and thus reduce pollution by cars, is not an example of a manager committed to the environment. Divergence of statements and actions are a sign of a threatened company, whereas a company's credibility is highly valued by staff, consumers and community. This applies also to operations in different countries. Multinational companies are tempted to adopt double standards in their subsidiaries in industrialized and in developing countries. Though companies are now detecting that these strategies are not particularly helpful in the long run, they are still being applied (Leonard, 1988).

Skill level of staff

It is not enough to be committed to the environment; a company needs to respond efficiently to the environmental challenge. Tougher environmental standards may require the installation and running of complex pollution abatement equipment, processes may have to be run with smaller tolerances and emissions will need to be analysed and monitored. Changes of products and processes will occur more often. The Green growth company will therefore be characterized by well-trained staff at all levels.

Research and development capacity

Green growth companies have demonstrated that they are able to anticipate or react quickly to changes in the market and environmental regulations.

Their growth is due to creativity in developing new products or modifying existing ones. Phosphate-free detergents, waste management services, modified chemical formulations and processes, design of recycling systems and alternative packaging are all a result of successful research and development (R&D). Companies which have short development cycles, creative, flexible R&D teams, and have access to information on clean technologies score high in our Environmental Challenge Scan.

If your company has developed specific know-how, for instance, on how to substitute or recycle materials, and/or to reduce pollution or wastes, the selling of this knowledge as a consultancy service to other companies can be a new profitable area of business.

Capital

It may take time before investments in "environment friendly" products and processes pay off. Companies may feel threatened by the fact that they will have to put money into products and processes without being able to clearly define the pay-back periods of these investments. Other investments made in the past may perhaps be written off in compliance with tougher environmental standards, rendering them unprofitable.

A Green growth company will have enough capital to enter into the Green venture or will have banks which are convinced that this company will be successful, despite or because of Green consumer pressure and environmental regulations.

Larger companies may have sufficient financial resources for expensive retrofitting in order to comply with environmental regulations, or may even discontinue a product. These companies might use their lobbying power to "soften" environmental regulations, or use the argument of job losses to receive grace periods for environmentally harmful processes and products.

Scoring your company

How does your company score in the Environmental Challenge Scan? If it scores high in all areas, you may wish to continue reading this book only to fine-tune your environmental management. You could benefit from

---

Signs of unsustainability

Concerning the gravity of environmental threats, the Brundtland report states: "Out of India's 3,119 towns and cities, only 209 had partial and only eight had full sewage treatment facilities. On the river Ganges, 114 cities each with 50,000 or more inhabitants dump untreated sewage into the river every day. DDT factories, tanneries, paper and pulp mills, petrochemical and fertilizer complexes, rubber factories, and a host of others, use the river to get rid of their wastes. The Hoogly estuary (near Calcutta) is choked with untreated industrial wastes from more than 150 major factories around Calcutta. Sixty per cent of Calcutta's population suffer from pneumonia, bronchitis, and other respiratory diseases related to air pollution".

Source: WCED, 1987, p. 240.

---

carrying out an environmental SWOT-analysis as described in section 2.1, designed to evaluate your Strengths, Weaknesses, Opportunities and Threats in order to consolidate your market position or enable you to enter new markets. If there are areas in which your company does not score well, you may wish to read the respective sections of this book. If you are managing a "threatened company", it would be useful to go through the whole book and establish an action plan as described in section 3.1.

Scoring the Environmental Challenge Scan reveals that there are internal factors which can be influenced by management and external factors, such as the environmental consciousness of a society and environmental standards, which at first glance seem to be out of management control. In this book we will demonstrate that business can influence environmental consciousness, acting on its own or within the framework of a business organization (see Chapter 4).

To devise mid-term and long-term business strategies, a manager has to predict how environmental consciousness and standards will develop, and how they could influence the company. Management will have to understand the reasoning of the discussions on sustainable development and the consequences for businesses in industrialized and developing countries. These issues will be discussed in the following sections.

1.2. Adjustment to sustainable development

Seveso, Bhopal, *Exxon Valdez* and Chernobyl, global warming and ozone depletion open our minds to the fact that there is something wrong with our development, but there are also minor incidents which may induce a change of mind: for instance, a manager who has to travel to work by public transport because a smog alarm prevents the use of private cars, or whose refuse is not collected because the city has nowhere to put it, will begin to contemplate the consequences if the outcome of rapid development is not properly taken into account. Since 1972, when the Club of Rome published its report *The limits to growth*, much has happened to our ecological environment and our perception of it. On the one hand, resources continue to be depleted and to be used in an inefficient manner. Irreversible environmental degradation can increasingly be found in all continents and we are far from sustainable development. On the other hand, environmental consciousness is increasing worldwide. How has the debate on economic growth and development evolved in the last 25 years?

The limits to growth states that if current levels of increase in population, industrialization, pollution, resource consumption and food production are maintained, the absolute limits to growth will be reached in the next hundred years. By changing the trend of growth it is, however, possible to reach an ecological and economic balance. In 1980, the United States Presidential Commission presented its report, *Global 2000*, in which the long-run implications of failure to act are spelled out in great detail. In 1983, the Commission chaired by the former German chancellor Willy Brandt linked development and environment. Its report, *Common crisis: Cooperation for world recovery*, makes recommendations for improved North-South cooperation regarding finance, trade, food, energy and the negotiation process. In 1987, the Brundtland Commission published its influential report, *Our common future*, in which the term "sustainable development" was coined. The Commission calls for

Table 2. Distribution of world consumption: Averages for 1980-82

Commodity	Units of per capita consumption	Developed countries (26% of population)		Developing countries (74% of population)	
		Share in world consumption (%)	Per capita	Share in world consumption (%)	Per capita
Food:					
Calories	Kcal/day	34	3395	66	2389
Protein	gm/day	38	99	62	58
Fat	gm/day	53	127	47	40
Paper	kg/year	85	123	15	8
Steel	kg/year	79	455	21	43
Other metals	kg/year	86	26	14	2
Commercial energy	mtce/year	80	5.8	20	0.5

Sources: WCED estimates based on country-level data from FAO, UN Statistical Office, UNCTAD and American Metal Association; WCED, 1987, p. 33, by permission of Oxford University Press.

action regarding population and human resources, food security, species and ecosystems, energy, industrial production and urbanization. The report became the basis for the Rio Conference. In 1992, Meadows and Randers, the authors of the 1972 Club of Rome study, reviewed their predictions. In *Beyond the limits* they argue that the limits of resource consumption and irreversible environmental degradation have already been reached and that change processes have become more costly, urgent and difficult. In the same year the United Nations Conference on Environment and Development (Rio de Janeiro) adopted Agenda 21, which proposes actions towards worldwide sustainable development (see box on page 19). In 1995, the World Trade Organization (WTO) was created. One of its major priorities is to deal with the issue of environment and trade.

This review of the environmental debate shows that over the last 25 years environmental issues have become a question of development and economic policies, where interests differ worldwide among and between industrialized, newly industrializing and developing countries. This becomes even clearer if we look into the worldwide distribution of resources, consumption and environmental degradation.

We are consuming non-renewable world resources, even though we are using most of them more productively than ten years ago. Wastes are left to later generations. The greenhouse effect, ozone depletion and acid rain change our whole ecosystem and challenge the survival of future generations.

There seems to be a large consensus that the present way of life in developed countries may not be extended to developing countries without causing an ecological collapse of our planet. During the early 1980s developed countries consumed more than ten times the commercial energy, steel and other resources, per capita, as did developing countries. Table 2, compiled by the World Commission on Environment and Development, gives a clear picture of the imbalance. There are disparities not only in consumption but also in worldwide pollution. In the case of greenhouse gases in 1987, just five countries (United States, USSR, Brazil, China and India) contributed 50 per cent of the warming potential added to the atmosphere that year (World Resources Institute, 1990).

Environmental degradation can be attributed to three main factors, namely population, per capita consumption of resources, and technology (cf. Ehrlich and Ehrlich, 1991). This can be formulated in the following manner:

Environmental load = Population × Per capita consumption
× Technology factor

The technology factor contains an efficiency factor and an emission factor. The efficiency factor determines how effectively resources are used, and the emission factor provides an indicator of emission per unit of output. To describe the local, regional and global stress on ecosystems, we distinguish between sources of environmental load and sinks. "Sources" play an opposite role to "sinks": while sources generate environmental pollution, sinks digest or absorb this pollution, for example tropical rain forests with their capacity to absorb CO_2.

These relationships, on the one hand, provide a basis for decisions towards sustainable development, while on the other hand they illustrate the associated conflicts. Whereas the major sources of environmental loads are located in industrialized countries, developing and newly industrializing economies are the ones which tend to provide the sinks. While industrialized countries argue that population control in the Third World is important, developing countries argue that per capita consumption has to be reduced in industrialized countries. More efficient technology is often offset by even higher consumption, e.g. in the case of energy in industrialized countries.

Measures towards sustainability are basically twofold. Firstly, environmental loads can be reduced by acting on population, per capita consumption and technology. Secondly, the absorptive and regenerative capacity of ecosystems can be increased.

The growing realization by world leaders that prosperity and well-being depend upon the rational use of the world's finite resources has marked a turning-point in attitudes. Most countries have overcome their suspicion of environmental protection as hindering development and reducing international competitiveness, and now see the need for resources, technology and expertise to be utilized to pursue environmentally sound development strategies. A new development problem and guiding principle for national and international industrial policies is emerging: sustainable development.

In its report (also known as the Brundtland report) the World Commission on Environment and Development (WCED) defines sustainable development as "development that meets the needs of the present without compromising the ability of future generations to meet their own needs" (WCED, 1987, p. 43). The Brundtland report calls for a (re)definition of the goals of economic and social development in terms of sustainability. According to the report, the required reorientation must consider the following topics:

☐ reviving growth;

☐ changing the quality of growth;

☐ meeting essential needs for jobs, food, energy, water and sanitation;

☐ ensuring a sustainable level of population;

☐ conserving and enhancing the resource base;

☐ reorienting technology and managing risk; and

☐ merging environment and economics in decision-making.

What will be the implications for industry of the proposed adjustment to sustainable development?

Issues addressed by Agenda 21

Agenda 21, adopted at the Earth Summit in Rio de Janeiro, reflects a global consensus and political commitment at the highest level on development and environment cooperation. The Agenda deals with both the pressing problems of today and the need to prepare for the challenges of the next century. The only way to assure ourselves of a safer, more prosperous future is to deal with environment and development issues together in a balanced manner. We must fulfil basic human needs, improve living standards for all and better protect and manage ecosystems. No nation can secure its future alone, but together we can – in a global partnership for sustainable development.

Section One: Social and Economic Dimensions
International cooperation
Combating poverty
Changing consumption patterns
Population and sustainability
Protecting and promoting human health
Sustainable human settlements
Making decisions for sustainable development

Section Two: Conservation and Management of Resources
Protecting the atmosphere
Managing land sustainably
Combating deforestation
Combating desertification and drought
Sustainable mountain development
Conservation of biological diversity
Management of biotechnology
Protecting and managing the oceans
Protecting and managing fresh water
Safer use of toxic chemicals
Managing hazardous wastes
Managing solid wastes and sewage
Managing radioactive wastes

Section Three: Strengthening the Role of Major Groups
Preamble to strengthening the role of major groups
Women in sustainable development
Children and youth in sustainable development
Strengthening the role of indigenous peoples
Partnerships with NGOs
Local authorities
Workers and trade unions
Business and industry
Scientists and technologists
Strengthening the role of farmers

Section Four: Means of Implementation
Financing sustainable development
Technology transfer
Science for sustainable development
Education, training and public awareness
Creating capacity for sustainable development
Organizing for sustainable development
International law
Information for decision-making

Source: Based on Keating, 1993.

Changing industrial structures

Sustainable development implies that industrial sectors, the products or processes of which are major polluters and energy intensive, which use scarce, nonrenewable materials and which cause hazardous waste will be restricted in their development or will have to pay higher costs for their resources and the environmental damage caused (Polluter Pays Principle).

These sectors are most likely to decline if they are not able to innovate, but if they change they could become growth sectors. Chemical companies are positive examples in this respect. New industrial growth sectors to emerge are the selling of cleaner and energy-saving technologies and the related know-how of water purification, handling and processing wastes, or the development of materials.

Adjustment to sustainable development cannot be separated from structural adjustment, undertaken by a number of countries to adapt their economies to market conditions and increased international competition (Jähnicke et al., 1989a; Simonis, 1989). We have been observing in many eastern European countries that the process of structural adjustment led especially to a "knock-out" to major polluters who are no longer profitable under market conditions and would not have survived the enforcement of environmental standards (Jähnicke et al., 1989b). A similar observation can be made for some state-owned industries in developing countries. Interrelated with changing industrial structures are a number of further implications of adjustment to sustainable development, which will be discussed in the following sections.

Environment-related business opportunities

☐ Services

Water management and purification; waste management; recycling; conservation of threatened or damaged habitats; technological research and development; training and advice for operators of pollution control in industry and public agencies; public transport.

☐ Production of equipment

End-of-pipe filters, scrubbers, treatment plants; equipment for the collection and transport of waste; monitoring and control equipment; public works materials; planning, engineering and design services.

☐ Environmentally friendly goods

Chloro-fluorocarbon (CFC) substitutes, biodegradable plastics, biodegradable detergents, catalysers, lead-free petrol, non-toxic paints, electric cars, solar energy for heating.

Source: Adapted from Commission of the European Communities, 1990, p. 81.

Relocation of enterprises

Industrial policies for sustainable development, and differing environmental standards from country to country (see section 1.3), are an argument for relocating production capacities, especially for pollution-intensive industries. Less rigid environmental regulations seem to give some developing countries a "comparative advantage" in the production of pollution-intensive goods, leading to further degradation of their environment. Industries that are most likely to move abroad to avoid environmental regulations are either declining industries that manufacture products for which consumption is decreasing or industries at an advanced stage of the international product cycle, where competitiveness is determined more by direct production and shipping costs than by possession of technological advantage (Leonard, 1988, p. 114).

In 1989 the industries of developing countries exporting to the Organisation for Economic Co-operation and Development (OECD) members would have incurred direct pollution control costs of US$5.5 billion if they had been required to meet the environmental standards then prevailing in the United States, according to a study conducted for the Brundtland Commission. In this respect the Brundtland report states: "The fact that these costs remain hidden means that developing countries are able to attract more investment to export manufactured goods than they would under a more rigorous system of global environmental control" (WCED, 1987).

The concentration of industrial activities in a few industrial mega-centres owing to raw material, transport or energy availability, or simply because of the colonial heritage, creates immense local pollution problems in developing countries such as in Cubatão in the São Paulo area of Brazil, Mexico City or many other locations. Some countries have started to relocate pollution-intensive industries far away from urban centres as a means of alleviating the urban pollution problem. This measure does not, however, reduce but only relocates pollution.

Green trade barriers

Managers are already facing a new kind of trade barrier. Industrial policies and regulations in favour of sustainable development can constitute a trade barrier for products which do not fulfil the environmental standards of the importing country. Requirements for recycling of products or packaging materials may result in such a high level of costs that the advantage of low-cost production in a developing country is lost (see *The Economist*, 15 June 1991).

From the perspective of industrialized countries, companies that operate under less severe environmental regulations (e.g. in Central and Eastern European countries) are accused of eco-dumping. The WTO has recognized that trade liberalization may be hampered by environmental trade barriers and has established environment and trade as a major area of its work. The means of overcoming – at least partially – an environmental blockage of trade might be the new international standard ISO 14000 defining environmental management systems. As business partners all over the world are now asked to comply with ISO 9000 quality standards, the ISO 14000 which links environment and quality might become an effective means to ensure that products are manufactured worldwide respecting certain environmental conditions.

Employment

Environmental policies may have important employment implications. Directives that guide or oblige industries to pollute less may also induce job losses in some industries as a result of cost increases.

On the other hand, the manufacture and introduction of a new generation of clean technologies may create jobs in research and the environmental infrastructure. Sewage and water treatment plants, waste disposal plants, and environmental management and administration in firms and public agencies create a variety of highly skilled jobs. Clean-up and conservation activities, by nature highly labour-intensive, are also a potentially important source of job creation (ILO, 1989; see also Commission of the European Communities, 1990). For enterprises it might be difficult to hire people for those highly skilled jobs on the labour market. Consequently, in-service training and retraining programmes will have to be established (see section 3.6). Many of the above-mentioned impacts of sustainable development policies on industry will be felt only in the medium or long term.

Long-term versus short-term perspectives

A manager, however, who is now responsible to the shareholders or owners of a company, could be inclined to disregard the medium and long-term consequences of the company's activities, particularly if there is a danger of a take-over bid from a hostile predator. As a taxpayer the company already shares, to some extent, the costs of sewage, waste treatment and pollution control at local, regional, national and international levels. If the company is a main polluter this policy may still pay off but, in any case, tougher regulations will have to be faced, especially as the concept of internalization of environmental costs is increasingly applied. The required retrofitting will most likely be more expensive than preventive measures would have been. Those working for environment-friendly companies would benefit the most if the "polluter pays" principle is applied: these companies would not then have to pay for the environmental damage caused by others.

As more and more companies are adhering to the new paradigm of "clean business", pressure on the major polluters by fellow industrialists, consumers, environmentalist groups and governments is mounting. The insight is growing that we cannot afford not to change. One of the many proofs is the increasing number of publications and well-attended seminars, conferences and workshops on "How to make your business lean, Green and clean". New rules of environmental business management are emerging, as the book *Changing course*, by the leading Swiss industrialist Stephan Schmidheiny (1992) demonstrates.

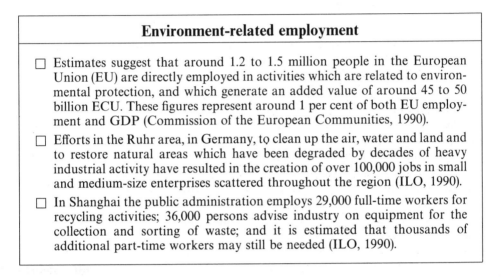

Environment-related employment

☐ Estimates suggest that around 1.2 to 1.5 million people in the European Union (EU) are directly employed in activities which are related to environmental protection, and which generate an added value of around 45 to 50 billion ECU. These figures represent around 1 per cent of both EU employment and GDP (Commission of the European Communities, 1990).

☐ Efforts in the Ruhr area, in Germany, to clean up the air, water and land and to restore natural areas which have been degraded by decades of heavy industrial activity have resulted in the creation of over 100,000 jobs in small and medium-size enterprises scattered throughout the region (ILO, 1990).

☐ In Shanghai the public administration employs 29,000 full-time workers for recycling activities; 36,000 persons advise industry on equipment for the collection and sorting of waste; and it is estimated that thousands of additional part-time workers may still be needed (ILO, 1990).

1.3. The regulatory framework: Economic incentives versus enforced actions

The legal and regulatory framework within particular countries will affect the role of a manager in preparing and implementing environmental decisions and actions. In countries where environmental agencies and policies have been established for the protection and enhancement of the environment, standards may well be available against which environmental management decisions can be judged or tested. Policies, details and the level of expertise within these agencies will vary, but in many cases a large company will possess more knowledge about the likely environmental effects of its proposed actions than the environmental authorities. A well-developed system of regulatory emission standards or objectives may place constraints on the environmental management of an organization, but will also provide a basis for measuring performance. In this context, the role of environmental management, articulated through senior management, is to determine the conditions which will apply to any given project and to ensure compliance with the regulations at an early stage. In this way the need to add expensive pollution control devices at the end of the construction period can be avoided.

What type of regulatory frameworks are managers facing or are likely to face in the future? Environmental regulations are a means towards three basic objectives:

☐ To protect and conserve the environment as well as maintain environmental quality (air, water and land quality).

☐ To protect human health.

☐ To regulate resource consumption.

A complex array of national and international laws, conventions, standards, recommendations, resolutions and self-regulations of industry have been established over the years to support initiatives in these areas.

On the national level we may differentiate between three categories of regulations to be respected by enterprises.[1] First, there are the *basic or enabling regulations* which define the general objectives and the powers of the legislative and executive branches of government, and its various agencies, including the competence of the central environmental agency (which may be an inter-ministerial board, a ministry of environment, etc.) and of the local and regional authorities. In some cases these regulations also establish a fiscal mechanism to ensure the financing of environmental measures. The basic laws often provide for a consultative process on environmental legislation, involving representatives of economic and social interests, community-action groups and other non-governmental organizations.

The second and perhaps most complex category is the *environmental quality or anti-pollution regulations*. Usually divided between air quality, water, marine pollution, solid waste and toxic materials regulations, they establish the quality criteria, define pollutants, set permissible limits and regulate control methods. These regulations ideally reflect the current "state of the art" and technology available for pollution monitoring and abatement. Since the scientific knowledge of the effects of pollutants on health and the technologies of prevention and abatement are in rapid evolution, the regulations need to be flexible enough to permit the necessary adjustments, without causing disruption of production and other economic activities. This flexibility can be obtained by referring in the law to "best available technology" or by incorporating the permissible limits into separate regulations which can be amended as needed by decree, without a long legislative process. Quality standards should not only ensure safety and health, but they should also be cost-effective. This is a main concern for managers who have to comply with these regulations.

---

[1] The following is based on Evan-Stein, 1990.

To illustrate this point, let us consider the management of air quality in an urban-industrial area. There are two basic types of air quality criteria: ambient air quality standards and pollutant emission standards. The former determine the maximum permissible level of pollutants in the ambient air. This means that in a less polluted area, more emissions are permitted. The latter establish the maximum amount of a pollutant which may be emitted from a fixed source (e.g. a stack) or a mobile source (e.g. car exhaust). Emission standards depend on the state of the technology, not of the environment. They are therefore less cost-effective. On the other hand, they are much easier to apply and to control, especially for the major sources (industry, power plants, etc.), whose emissions can be monitored with relative ease. Motor vehicle emissions can also be controlled by periodic inspection.

Ambient air quality standards, on the other hand, require a very elaborate network of automated monitoring stations, including meteorological parameters. If properly designed, these standards can be more flexible and less costly to implement than emission standards. In practice a combination of both standards is often applied.

The third category of environmental regulations is concerned with *resource conservation or resource management*. The oldest and most widely applied are land use regulations, but other scarce resources are regulated in many countries, such as water, forests, minerals and nature sites. These laws have the objective of allocating the resources to satisfy various needs, while providing for their long-term availability through conservation and restoration. New planning techniques, such as resource inventories, environmental impact analysis and the establishment of master plans are becoming part of the regulatory framework. Resource management laws have an important influence on managerial planning and economic value and are indispensable for production.

In many countries *specific economic instruments* are an integral part of the environmental legislation. Their aim is to discourage business activities which harm or make excessive use of resources by rendering these activities unprofitable. The most common instruments are charges, subsidies, deposit/refund systems and trading licences, as well as enforcement incentives (Opschoor and Vos, 1989).

A number of countries are currently discussing the pros and cons of reforming their tax system. These reforms aim at taxing energy consumption and CO_2 production and levying less tax on human labour. In Germany, for example, Greenpeace financed a study coedited by one of the leading economic institutes of the country which showed that a tax system based on a CO_2 tax would be economically feasible and would contribute to the creation of jobs. It is clear, however, that this would lead to substantial competitive disadvantages for industrial sectors that consume huge amounts of energy, such as the chemical industry. That is why industry prefers voluntary agreements that take into account the specific problems and the potential for solutions of different industrial sectors.

A widely discussed instrument is the creation of market mechanisms, allowing companies to buy "rights" to actual or potential pollution, or where they can sell their "pollution rights" or their process residuals. Several forms exist, among which the emissions trading systems are the most important.

Improvement of the working environment is a good example of combining both the economic and environmental interests of an enterprise. Healthy employees working in pleasant, clean and hazard-free surroundings are likely to be more productive than an unhealthy workforce operating in unhygienic conditions. Regulations controlling working conditions are now well established in most countries and are regularly being extended and refined. These aspects will be dealt with in more detail in section 2.9.

Voluntary agreements between industry and government as well as *self-regulation* of business sectors are another cost-effective instrument of environmental protection. Once an agreement has been reached on an environmental quality target, voluntary agreements and self-regulation allow business to look for the most cost-effective way to achieve this target without having a regulatory mechanism imposed by government. By actively proposing such arrangements to central or local government, individual enterprises or business associations can shape environmental regulations to protect the environment and suit their interests. Managers will have to be aware that these voluntary agreements and self-regulations will be denounced by the environmentally aware public if they are too much in favour of enterprise interests. This would result in a loss of confidence in the commitment of industry.

As pollution, like many business activities, does not stop at country borders, a range of *international regulations* have been established. It is a quickly changing regulatory framework difficult to keep up with. There are three main categories of international environmental agreements:

☐ Conventions on transboundary pollution (e.g. the 1979 Conventions on Long-Range Transboundary Air Pollution and its Helsinki Protocol, providing for a 30 per cent reduction of sulphur emissions from 1980 levels to be achieved by 1993 at the latest).

☐ Conventions on resources shared between two or more States in a region (e.g. the UNEP's Regional Seas Convention).

☐ Conventions on the use of resources of the "global commons", i.e. oceans, atmosphere and outer space (e.g. the Montreal Protocol on Substances that Deplete the Ozone Layer and reducing'CFC production).

The United Nations Environment Programme (UNEP) publishes a list of international agreements for each session of the General Council (e.g. UNEP, 1989a) in the field of the environment, which can assist managers in finding their way through the regulatory jungle. A selected number of these agreements have been listed in table 3.

International environmental law obliges signatory governments to enact appropriate laws within their national jurisdiction. In fact, some

Table 3. International agreements in the field of the environment

In 1989 the UNEP Register of International Treaties and other agreements already had 140 entries. In the following, selected agreements relevant to business are listed:

General environmental concerns

UNEP Principles of Conduct in the Field of the Environment for the Guidance of States in the Conservation and Harmonious Utilization of Natural Resources Shared by Two or More States (Nairobi, 19 May 1978)

UNEP Goals and Principles of Environmental Impact Assessment (Nairobi, 17 June 1987)

Convention on Environmental Impact Assessment in a Transboundary Context (Espoo, 25 February 1991)

Nature and living resources

International Convention for the Regulation of Whaling (Washington, DC, 2 December 1946) as amended

Convention on International Trade in Endangered Species of Wild Fauna and Flora (Washington, DC, 3 March 1973) as amended

International Tropical Timber Agreement (Geneva, 18 November 1983)

Atmosphere

United Nations/ECE regulations concerning gaseous pollutant emissions from motor vehicles, pursuant to the 1958 agreement concerning the Adoption of Uniform Conditions of Approval and Reciprocal Recognition of Approval for Motor Vehicle Equipment and Parts (Geneva, 1 August 1970, as amended and supplemented to 1991)

Convention on Long-range Transboundary Air Pollution (Geneva, 13 November 1979)

Vienna Convention for the Protection of the Ozone Layer (Vienna, 22 March 1985), and Montreal Protocol (16 September 1987)

Marine environment

International Convention for the Prevention of Pollution of the Sea by Oil (London, 12 May 1954) as amended on 11 April 1962 and 21 October 1989

International Maritime Dangerous Goods Code (London, 16 September 1965) as amended

International Convention on Civil Liability for Oil Pollution Damage (Brussels, 29 November 1969) and related Protocol as amended

Convention on the Prevention of Marine Pollution by Dumping of Wastes and Other Matter (London, 29 December 1972) as amended (Mexico City, Moscow, [Washington, DC])

Convention on the Prevention of Marine Pollution from Land-based Sources (Paris, 4 June 1974) as amended on 26 March 1986

Various Regional Conventions for Co-operation on the Protection of the Marine Environment from Pollution

UNEP Guidelines Concerning the Environment Related to Offshore Mining and Drilling within the Limits of National Jurisdiction (Nairobi, 31 May 1982)

United Nations Convention on the Law of the Sea (Montego Bay, 10 December 1982)

International Convention on Oil Pollution Preparedness, Response and Co-operation (London, 29 November 1990)

Hazardous substances and processes

European Agreement Concerning the International Carriage of Dangerous Goods by Road (Geneva, 30 September 1957) as amended

Table 3. *(cont.)*

Vienna Convention on Civil Liability for Nuclear Damage (Vienna, 21 May 1963) and related protocol

European Agreement on the Restriction of the Use of Certain Detergents in Washing and Cleaning Products (Strasbourg, 16 September 1968)

OECD Procedures and Requirements for Anticipating the Effects of Chemicals on Man and in the Environment (Paris, 7 July 1977)

OECD Decisions on the Mutual Acceptance of Data in the Assessment of Chemicals (Paris, 12 May 1981) and related recommendations

FAO International Code of Conduct on the Distribution and Use of Pesticides (Rome, 19 November 1985) as amended

UNEP London Guidelines for the Exchange of Information on Chemicals in International Trade (Nairobi, 17 June 1987) as amended

UNEP Cairo Guidelines and Principles for the Environmentally Sound Management of Hazardous Wastes (Nairobi, 17 June 1987)

Basel Convention on the Control of Transboundary Movements of Hazardous Wastes and their Disposal (Basel, 22 March 1989)

Working environment (see box in section 2.9, on p. 91)

countries have set national objectives well ahead of the time frame of the international conventions. The United States intends to reduce SO_x by 50 per cent by the year 2000. Developed countries, as per the Montreal Protocol, have already phased out new production of halons, CFCs, carbon tetrachloride and methyl chloroform, except for small quantities required for essential uses.

Even though international environmental agreements are contributing to the harmonization of environmental standards of different countries, there are still enormous divergencies which can act as trade barriers and influence the competitive position of these countries. Thus, it is not surprising that one of the major issues of European integration, for example, is the harmonization of environmental policy measures. The variety of measures taken by the European Union is shown in table 4.

Differing environmental standards between countries are becoming an increasing problem for managers:

☐ Companies from industrialized countries will have to decide (or will be forced to decide by the Green movement) if they want to build their profitability on the comparative advantage of lower environmental standards in developing countries.

☐ Companies from developing countries might find an increasing number of trade barriers as their products do not meet the rigid environmental standards of industrialized countries.

☐ Recycling regulations may force producers to take used products back and dispose of them in their country of origin.

☐ Product liability regulations vary greatly from country to country. Products which are banned in one country, such as pesticides, are still allowed

Table 4. Environmental policy measures taken by the European Union

	Air	Noise	Water	Waste	Chemicals	Nature protection
Trading licence				■		
Installation standards	■		■			
Product standards					■	
Environmental impact assessment	■	■	■	■	■	■
Production emission standards	■	■	■	■	■	
Production limits	■					
Trading limits	■			■	■	■
Emission standards for installations	■	■	■	■		
Standards for specific areas	■		■			■
Voluntary agreements				○	■	
Product levies				○		

○ = optional to the member States.

Source: Commission of the European Communities, 1990, p. 80.

in other countries. In addition, a producer may be liable for future damage caused by a product considered "harmless" today.

No less important than the regulatory provisions themselves are the institutional arrangements for their administration (Evan-Stein, 1990; UNEP/IEO, 1992). One key aspect, from the viewpoint of the enterprise, is the question of competence: their concentration in the hands of a single authority, or their division between many authorities. Owing to the complexity and multidisciplinary nature of environmental management, there is a tendency to disperse the application of environmental laws between various sectoral ministries (transport, industry, agriculture, energy, etc.). In the absence of a strong central authority, the resulting multiplicity of standards and procedures and the occasional overlapping and conflicting skills can be the cause of much delay and uncertainty, as well as added costs to enterprises and their management.

On the other hand, too much centralization of authority can be the cause of delay, since the day-to-day control and management of the environment cannot be efficiently conducted from a distance. Most countries have therefore developed a structure of coordination and delegation of authority between the central environment agency, the sectoral ministries responsible for economic affairs, health, etc., and the local authorities, empowering the central environment agency to establish basic standards, provide information and support, and act as "watchdog" while sectoral and local authorities directly administer the regulations.

Industry in most cases will be interested in having enforcement systems of environmental regulations which function efficiently, as unenforced regulations are a potential threat which could require high investments and quick reactions when they are suddenly enforced.

In this respect, management should collaborate with the relevant authorities to develop a predictable, coherent, market-oriented and well-enforced system of environmental regulations as the basis for an environmental management which is able to turn threats into opportunities. How to do this will be outlined in the following section.

Monitoring compliance with environmental standards

☐ All necessary permits are up to date, and reflect actual operations, and copies are available.

☐ Where there are no regulatory requirements, appropriate written standards have been developed internally.

☐ All emissions monitoring required by the regulatory authorities is carried out using satisfactory procedures.

☐ Where there are no regulatory requirements for emissions monitoring, there is a means of regular feedback on the efficiency of environmental protection measures.

☐ Appropriate environmental monitoring is carried out according to statutory requirements or other defined objectives and uses satisfactory procedures.

☐ Records of all monitoring data are available and have been reported to regulatory authorities as required.

☐ Steps are taken to identify the causes and take remedial action when:
  (i) monitoring data or other information indicate non-compliance;
  (ii) receiving environment monitoring indicates unacceptable environmental degradation.

Source: Adapted from UNEP/IEO, 1990b.

1.4. The management challenge: Converting threats into opportunities

"Companies that take the environment seriously find themselves changing not only their processes and their products, but also the way they run themselves. Often, such changes go hand in hand with improvements in the general quality of management. Badly managed companies are rarely kind to the environment; conversely, the companies who try hardest to reduce the damage they do to the environment are usually well managed", states the British business journal *The Economist* (8 September 1990, p. 22) in its environmental survey. This is supported by Fritz (1995), who found for German enterprises that those enterprises which are leaders in environmental protection are also economically successful. He therefore argues that environmental protection and economic success are in fact complementary, and not contrary to each other.

The Environmental Challenge Scan, introduced in section 1.1, supports this finding by attributing the best chances of Green growth to those companies which have an open-minded, committed and highly skilled management and staff. They are creative and have short development cycles with which to adapt flexibly to changing markets, and are able to convince their banks of their future growth potential in a Green market.

To look more closely into what makes companies successful in turning threats into Green growth opportunities, we will extract a number of principles which describe what environmental management is. The explanation of these principles serves also to summarize the following chapters of the book and assists managers in selecting the parts to read that interest them the most.

Those principles have been well defined in the "Business Charter for Sustainable Development", adopted by the International Chamber of Commerce (ICC) in November 1990. The full text of the charter can be found in Appendix 4 of this book.

In the following, the principles of environmental management are explained:

1. *Corporate priority.* **To recognize environmental management as among the highest corporate priorities and as a key determinant to sustainable development; to establish policies, programmes and practices for conducting operations in an environmentally sound manner.**

The inclusion of environmental protection in the corporate mission statement will be the first step to publicizing the commitment of an enterprise, which will have to be followed by concrete actions endorsed by the owners and managers of the company. The important role of credible top management will be stressed in section 2.1. Environmental protection cannot become a corporate priority if employees do not believe in it. Communication of the company objectives, programmes and practices to the employees, as well as their participation, is an essential part of successful environmental management (see section 3.5).

The importance of commitment and skills

The American company Geneva Steel is one of the few steel makers in the world with a good environmental reputation – and one of the world's most profitable. Its chairman began his career as a regulator in the United States Environmental Protection Agency. He therefore understands better than most industrialists how regulators work.

At the age of 40 he is a good decade younger than most of his fellow steel bosses. Above all, his Utah plant employs a largely Mormon workforce with an average of one college year of education.

An intelligent workforce has allowed him to solve one of the most intractable problems of a steel mill and cut emissions from his coke ovens to less than a quarter of the permitted maximum. The secret is simply to take care of the oven doors. The fact that he is known to care about environmental problems encourages employees to come up with Green ideas of their own. And because the company is profitable and he has a controlling stake in it, he has been able to take large investments in cleaner technology well before the state's regulations would have required it.

Source: *The Economist*, 8 Sep. 1990, p. 22.

---

2. *Integrated management.* **To integrate these policies, programmes and practices fully into each business as an essential element of management in all its functions.**

Each management function within an enterprise is responsible for the implementation of an environmental policy. Production management must ensure that operations are efficient and as non-polluting as possible (see section 2.5). Project management must ensure that projects are harmoniously integrated into the physical and social environment (see section 2.6). Marketing management must ensure that products sold are clean, safe and efficient and meet consumers' needs (see section 2.2). Personnel management must ensure that all workers are aware of the need to protect the environment (see section 2.8). R&D management must develop new products and processes which are non-polluting, safe and efficient (see section 2.3). Materials management provides recyclable and non-polluting materials (see section 2.4.) Financial management will assist in making pollution prevention pay (see section 2.7). Occupational health and safety measures will have to protect workers from adverse environmental conditions (see section 2.9). All these functions have to be integrated by the appropriate organizational structures and processes on all management levels (see section 3.4).

3. *Process of improvement.* **To continue to improve corporate policies, programmes and environmental performance, taking into account technical developments, scientific understanding, consumer needs and community expectations, with legal regulations as a starting-point; and to apply the same environmental criteria internationally.**

To be effective, the environmental management process must be systematic, detailed and integrated into all functional management decisions. Action plans should be set up to ensure continuous improvement of environmental performance of an enterprise (see section 3.1).

Conventionally, managers make decisions on the basis of internal costs and benefits to their organizations, whereas environmental management takes into account the external positive and negative environmental effects of actions that do not necessarily have a direct economic relationship with the enterprise or action proposed. Good environmental management can be of economic advantage to an organization, but its justification must be based on the recognition of these wider responsibilities. Creating this environmental awareness is the greatest challenge in developing environmental management in a company. Successful environmental management, just like the development of positive policies regarding equal opportunities, employment or the creation of better working conditions, requires the extensive participation of workers/employees and their representatives (see section 3.5).

Benefits of environmental management for a company

Economic benefits

☐ Cost savings:
- Savings due to reduced consumption of energy and other resources.
- Savings due to recycling, selling of by-products and wastes, resulting in decreased waste disposal costs.
- Reduced environmental charges, pollution penalties, compensations following legal damage suits.

☐ Revenue increases:
- Increased marginal contribution of "Green products" which sell at higher prices.
- Increased market share due to product innovation and less performative competitors.
- Completely new products open up new markets.
- Increased demand for a traditional product which contributes to pollution abatement.

Strategic benefits
- Improved public image.
- Renovation of product portfolio.
- Productivity improvement.
- Higher staff commitment and better labour relations.
- Creativity and openness to new challenges.
- Better relations with public authorities, community and Green activist groups.
- Assured access to foreign markets.
- Easier compliance with environmental standards.

The importance of environmental regulations as basic rules for environmental management has already been demonstrated in section 1.3. These are often a cause of differences in management practices with companies operating in both developed and developing countries. Distant subsidiaries may pay less heed to company policy and safety standards than those closer to headquarters. The desire to minimize costs may lead to double standards in operations in different countries and local staff may be less committed to corporate policy. Regular environmental audits can be an appropriate means to unify environmental management practices of a company (see section 3.3).

4. *Employee education.* **To educate, train and motivate employees to conduct their activities in an environmentally responsible manner.**

Education and training to raise awareness about environmental problems, to change attitudes and behaviour and to provide the necessary skills to act in an environmentally responsible manner are important components of environmental management which concern staff as well as management (see section 3.8.)

5. *Prior assessment.* **To assess environmental impacts before starting a new activity or project and before decommissioning a facility or leaving a site.**

For many companies, the turning-point in the introduction of environmental management came when enormous amounts of money were lost because they were not allowed to start new or extend existing facilities owing to public opposition and stricter authorization procedures prescribed by the government authorities.

Given this experience, it is clearly very risky to undertake, finance or approve a major project without first taking into account the environmental consequences (UNEP, 1988). Therefore, management should be familiar with the main features of an environmental impact assessment (EIA), which could be carried out either by specialists of the company or by outside consultants. The principles of EIA are discussed in section 3.2.

6. *Products and services.* **To develop and provide products or services that have no undue environmental impact and are safe in their intended use, that are efficient in their consumption of energy and natural resources, and that can be recycled, reused, or disposed of safely.**

Environmental management is often associated with cleaning-up production processes and not so much with producing environment-friendly products. As we have seen in the Environmental Challenge Scan, both products and processes determine the environmental performance of an enterprise. At the development stage, therefore, the whole life cycle of a product has to be considered (see section 2.3). Marketing management has to translate the consumers' desire for environment-friendly products into a product concept (see section 2.2). Materials management (see section 2.4) will have not only to purchase recyclable materials but also to handle wastes and take back used products and substances. Car companies, for example, are currently starting to scrap their cars in experimental plants.

Some chemical companies are offering to take back contaminated substances and clean them up.

This extended responsibility of producers for their products from cradle to grave is a significant distinguishing mark of environmental management compared with traditional "after sales" concepts. Environmental management means managing the product life cycle. This also leads to changing links between companies and their customers.

7. *Customer advice.* **To advise and, where relevant, educate customers, distributors and the public in the safe use, transportation, storage and disposal of products provided; and to apply similar considerations to the provision of services.**

The label "environment-friendly" has become a major advertising argument. This argument, however, is often misleading for consumers who do not know the basis on which a product has earned this label. In the light of the extended responsibility for its products, as described in the explanation of principle 6, companies share a responsibility to make consumers environmentally literate, to explain why products are "environment-friendly" and provide consumers with adequate information for the appropriate selection and use of products. The design of disposal systems for used products or packaging materials is increasingly becoming the responsibility of producers and distributors. Classic examples are the return of used motor oil containers or batteries. Environmental management also means developing an environment-conscious and literate sales and marketing staff (see section 2.2).

Turning threats into opportunities
Examples from developing countries

☐ India: Harihar Polyfibers in India has developed the technology to recycle process waste in the synthetic fibre industry and to recover materials such as caustic soda for reuse, with significant cost savings. The same company also achieved reduced oil consumption with heat control technology which improved product quality.

☐ United Republic of Tanzania: A Portland cement factory introduced a new system for the rehabilitation of electrostatic precipitators which was reported to produce a significant reduction of air pollution and produces cost savings of about US$8,000 per day.

☐ Botswana: When it experienced difficulty with the supply of materials, an enterprise developed a small-scale labour-intensive plant to produce moulded fibre newsprint from recycled newspaper and agro-waste, which had previously caused severe pollution.

☐ Zambia: The Ndola Lime Works introduced rotary kilns and technology changes in lime making, which led to improved output and efficiency, lower costs, cleaner working areas and fewer accidents.

Source: Winter et al., 1991, pp. 13 and 20.

Turning threats into opportunities
Examples from industrialized countries

- ☐ A German battery manufacturer introduced a new line of batteries without either mercury or cadmium, two chemicals which were both difficult to dispose of and toxic. Within six months its share of the US$120 million British supermarket business had jumped from 5 to 15 per cent (*Fortune*, 23 October 1989, p. 52).

- ☐ A French disposable nappy manufacturer, a subsidiary of a Swedish pulp and paper company, pioneered the use of nappies made from pulp that is bleached without using toxic chlorine gas. The move helped the company to increase its share of the more than US$500 million British market from 10 to 13 per cent (Tully, 1989, p. 52).

- ☐ A chemical plant formerly sent 6 million pounds of waste nylon to landfill sites each year at a cost of 2 cents a pound, or US$120,000 a year. This waste nylon is now recycled into engineering polymers, used to produce parts for office furniture and kitchen utensils (Newall, 1990, p. 92).

- ☐ A company which produced its polyethylene fencing entirely from recycled resins found that sales of the product subsequently boasted a 50 per cent increase because lower manufacturing costs resulted in a reduced selling price (Newall, 1990, p. 92).

- ☐ The fight against air pollution has produced clear benefits for a Swedish manufacturer of anti-air pollution gear for power plants. The company increased revenues from 1988 to 1989 by 50 per cent to US$700 million (Tully, 1989, p. 47).

---

8. *Facilities and operations.* **To develop, design and operate facilities and conduct activities taking into consideration the efficient use of energy and materials, the sustainable use of renewable resources, the minimization of adverse environmental impact and waste generation, and the safe and responsible disposal of residual wastes.**

This aspect is often the first to be considered by managers in the event they are asked to explain what environmental management means. This may be due to the fact that the development, design and operation of facilities are the functions within the enterprise which are most affected by environmental regulations. Many company programmes still reflect in their names the fact that they resulted from pollution prevention, abatement or waste minimization initiatives, for example, "Pollution Prevention Pays" of the 3M company, or "Save Money and Reduce Toxics" (SMART) of Chevron. Planning and running facilities and operations calls for integrated management as described under principle 2.

In the planning process of new or extended facilities and operations Environmental Impact Assessment (EIA) should be carried out (see section 3.2). Project management should include environmental considerations in all project phases (see section 2.6). Environmentally committed production management (see section 2.5) must ensure that operations comply with environmental regulations, respect occupational health and safety standards (see section 2.9) and make the best use of available resources (see sections 3.7 and 3.8). Regular environmental audits (see section 3.3) can help production management to assess and improve the environmental performance of operations. Employee participation schemes (see section 3.5) are particularly helpful towards this objective.

9. *Research.* **To conduct or support research on the environmental impacts of raw materials, products, processes, emissions and wastes associated with the enterprise and on the means of minimizing such adverse impacts.**

Research and development (R&D) creates a company's product portfolio for the future and plays a key role in converting cleaning-up operations into new business opportunities. This gives strategic importance to a creative R&D team and short development cycles to meet changing consumer demands and environmental standards in a flexible way. In this respect, the Environmental Challenge Scan considers such R&D capacity as a characteristic of a Green growth company. R&D carried out within the concept of environmental management aims at using renewable materials and reducing the consumption of resources, as well as minimizing pollution and wastes associated with the manufacture, storage, use and disposal of the product. How to include environmental considerations into R&D is dealt with in section 2.3.

10. *Precautionary approach.* **To modify the manufacture, marketing or use of products or services or the conduct of activities, consistent with scientific and technical understanding, to prevent serious or irreversible environmental degradation.**

Green growth companies have demonstrated that it often pays to be among the first to modify products and processes before consumer pressure mounts, before serious accidents happen or before environmental regulations have to be enforced. Enterprises taking a precautionary approach, moreover, will not start production of a new product or will discontinue an existing process if there are signs of negative health effects on workers or the general population, or if environmental damage is caused. Environmental management in this sense means to act voluntarily in a preventive way without being forced by environmental regulations (see section 2.1).

11. *Contractors and suppliers.* **To promote the adoption of these principles by contractors acting on behalf of the enterprise, encouraging and, where appropriate, requiring improvement in their practices to make them consistent with those of the enterprise; and to encourage the wider adoption of these principles by suppliers.**

Traditionally, companies have always looked for suppliers who could provide parts or components at the lowest price while meeting the company specifications. The purchasing company was eager to exploit competition among suppliers internationally by global sourcing and did not bother how, and under what conditions, components and parts were manufactured. Even though this is still the case for many companies, a new pattern of purchaser/supplier relations is emerging, which is already quite common in the automobile industry. The purchasing company and suppliers cooperate in all stages of the development, design and manufacture of parts and components. The supplier becomes a "family member" of a business group and, consequently, will have to share the corporate identity and values of the business group (Womack et al., 1990).

Such a close exchange between supplier and purchaser leads to similar standards of operation, which is also desirable for the attainment of environmental standards. Environmental management has to ensure that not only the supplier but also contractors, and franchising or licence partners achieve comparable environmental performance as the purchasing or out-contracting company (see section 2.4).

12. *Emergency preparedness.* **To develop and maintain, where significant hazards exist, emergency preparedness plans in conjunction with the emergency services, relevant authorities and the local community, recognizing potential transboundary impacts.**

An emergency or near emergency was, for a significant number of companies, the starting-point for their environmental management thinking. Green pressure groups and communities have become very sensitive to the inherent dangers of local industry. There is a certain mistrust on the part of the community which gives rise to fears concerning the extent to which it might be negatively affected by leaks, spills, explosions or fire. An environmentally conscious company not only needs to take all internal precautions to avoid emergencies but also has to make the community and responsible authorities aware of hazards. The company, jointly with the community, should develop an emergency response plan and train the local residents on how to act in the event of an emergency. Where regional or transboundary impacts may be caused, e.g. by the release of toxic substances into a major river, the respective emergency plans need to be established in collaboration with regional, national and international authorities (see sections 3.5 and 3.9).

13. *Transfer of technology.* **To contribute to the transfer of environmentally sound technology and management methods throughout the industrial and public sectors.**

Companies which comply with strict environmental standards, or have established environmentally sound working practices on a voluntary basis, will have an interest in other companies also adopting similar standards. Only then will the forerunners have an advantage in having been the first. Environmentally conscious companies may also be interested in changing the polluter image of their industrial branch and are therefore prepared to hand over knowledge to other companies in the sector. Many companies, especially in developing countries, may be willing to modify products and processes without having access to or being unable to manage the appropriate cleaner technologies. A number of environmentally committed companies and international institutes are now providing support information services on environmental management practices. These institutions include, for example, the International Environmental Bureau and the International Cleaner Production Information Clearing House (ICPIC) (see Chapter 4).

14. *Contributing to the common effort.* **To contribute to the development of public policy and to business, governmental and intergovernmental programmes and educational initiatives that will enhance environmental awareness and protection.**

15. *Open-mindedness to concerns.* **To foster open-mindedness and dialogue with employees and the public, anticipating and responding to their concerns about the potential hazards and impacts of operations, products, wastes or services, including those of transboundary or global significance.**

As both principles 14 and 15 are related, they will be discussed together. The combination of an informed public, "good business" practice and environmentally sound decisions will strengthen the case for environmental management. An informed public will also provide a framework for environmental management by setting the parameters of what is considered acceptable and the priority concerns of society.

Similarly, environmental pressure groups can impede sound environmental management by focusing attention on one issue at the expense of others. Environmental pressure groups assist the task of management, however, by highlighting environmental concerns and providing arguments to counter the case for growth or the creation of jobs at any cost. In this respect management must acquire communication skills to deal with the general public and pressure groups (see section 3.5).

An open dialogue not only with the outside world but also with employees is a major asset in environmental management. Employees working in a company with a polluter image will be less motivated to contribute to the company's success and to act as ambassadors in an environmentally conscious society than employees of a Green growth company. Labour relations of "dirty" companies seem to be more conflictive than those of less polluting companies (Gladwin, 1980).

Employer/employee dialogue on environmental issues should not be limited to company level, but should form part of a tripartite dialogue between employers' associations, trade unions and government. Chapter 4 deals with these aspects of environmental management in more detail.

Many companies express their concern for the environment by supporting community initiatives to preserve nature, for example, by donating trees, assisting in cleaning up polluted areas, financially assisting school environment programmes, donating wetlands adjacent to the plant to the regional conservation authority or sponsoring environmental awards. By these actions companies demonstrate that they and their employees form part of the community and that they are prepared to share responsibility for the environment.

16. *Compliance and reporting.* **To measure environmental performance; to conduct regular environmental audits and assessments of compliance with company requirements, legal requirements and these principles; and periodically to provide appropriate information to the Board of Directors, shareholders, employees, the authorities and the public.**

Environmental management is not a specific kind of philanthropy; it is a business strategy, the results of which have to be measured in economic and ecological terms. One way of measuring a company's environmental performance is by improving its regular environmental audits as explained in section 3.3. Small group activities such as quality circles and suggestion schemes contribute to setting goals at workplace or unit levels, the achievements of which can be directly assessed by the participating employees (see section 3.5). Financial management will have to assess if "pollution prevention pays" (see section 2.7).

While measuring environmental performance seems to be accepted as a necessary part of environmental management, companies are still reluctant to openly inform interested parties of the results. Even though companies may not be willing to release a detailed statement on environmental performance to the general public, they should have a confident relationship with employees and the relevant authorities. Experience has shown that hiding adverse environmental effects of a company's operations will lead to a loss of credibility in the eyes of the public. To establish and maintain a Green and clean company image, the importance of a credible communication strategy cannot be underestimated (see section 3.5).

The integration of these 16 principles of environmental management into the practical tasks of running a company on a day-to-day basis requires a long-term business perspective. Where to start, how to define priorities of action, and what are the right actions to undertake in order to turn threats into opportunities will be discussed in the following chapters.

2

The managerial response

This chapter explains:
- [ ] how to establish a corporate environmental policy; and
- [ ] how to integrate environmental considerations into all functional management areas.

2.1. Top management: From policies to action

In many boardrooms there have been discussions concerning the extent to which enterprises should commit themselves to environmental protection and publicize this decision. There are many good arguments for doing so:

- [ ] First take up the environmental challenge before your competitors do so.
- [ ] Be environmentally responsible and let it be known. Demonstrate to the outside world, governments, clients, investors and community that you take environmental issues seriously and that you practise environmental protection successfully.
- [ ] Make pollution prevention pay. Being considered an "environmentally friendly enterprise", especially if you exceed the required regulations, gives you an advantage in relation to clients, the community and regulatory bodies.
- [ ] Enhance staff commitment. As the general awareness of environmental issues is increasing, people do not want to work for an enterprise renowned as a major polluter. Recruitment, retainment and commitment depend largely on a positive company image.

There are indeed many starting-points to becoming Green. Often it is not top management which decrees radical change. Many companies have experienced a period of uneasiness about how to tackle environmental questions; middle management has taken up environmental initiatives, often without the explicit backing of top management until a certain maturity in environmental thinking in the company has been achieved.

Enterprises which experience direct pressure from clients such as retailers, consumer goods manufacturers, suppliers or service companies, or which have a polluter image such as those in the chemical industry, are those most likely to make a comprehensive change. In which part of the company change is made depends largely on individual commitment and resistance. If, for example, the head of the R&D department is committed to environmental protection, there is a likelihood that all efforts will be made to replace harmful materials with environmentally friendly ones, or to reduce the

---

Ten steps to environmental excellence

1. Develop and publish an environmental policy.
2. Set targets and continue to measure achievements.
3. Clearly define the environmental responsibilities of management and staff.
4. Communicate policy, targets and responsibilities to staff and the outside world.
5. Allocate adequate resources.
6. Educate and train management, staff, consumers and the community.
7. Monitor, audit and report.
8. Monitor the evolution of the Green debate.
9. Contribute to environmental community programmes, invest in environmental science and technology.
10. Help build bridges between the various interests.

Source: Based on Elkington and Burke, 1989.

---

energy consumption of a new generation of products. If this Green R&D manager has to face resistance from the materials management department, which feels that the new Green fashion has reduced its power and freedom of action, R&D will be in a difficult position. In such a constellation, each proposal for alternative materials put forward by R&D would be blocked by materials management referring to a higher price, less reliable suppliers, larger purchasing lead times and other excuses.

The type of ownership of a company also influences the direction the company takes to become Green. Short-term profit orientation counters profound change. An enterprise formed as a (consumer) cooperative, or ownership by a foundation, tends to be more open to pressures to become Green and lean. In Switzerland the largest retail (supermarket) chain is a consumer cooperative which has led the way in including environmental considerations in its business strategy and action. This is clearly visible in the products offered and how they are packed in shops all over the country.

A foundation associated with the cooperative conducts research into alternative forms of economic development.

Many companies would not like to go as far as this Swiss cooperative, at least not for the moment. Managers would like to know where to start and there is a persistent fear that once you have begun there is no stopping the environmental movement in the company. An evolution? A revolution? Wait and see? What is the right strategy for our company?

In defining its environmental strategy, a company will have to decide if environmental issues are a matter of compliance without obvious economic benefits or if the environmental debate can be used to develop competitive advantages. Drawing an analogy with Herzberg's Two-Factor Motivation theory, Wehrmeyer (1996) classifies environmental issues as issues of environmental hygiene and issues that are environmental motivators. According to Wehrmeyer, environmental hygiene factors could include pollution control limits, matters of environmental health and safety, waste management and the threat of environmental persecution. Environmental motivators could be growing market demand, the capability of a firm to gain or maintain competitiveness through environmental management, or the opportunities arising from developing new markets or growing within existing markets. Seaswift papermill in the United Kingdom, for example, decided deliberately that for their operations environmental issues are hygiene factors which obviously need addressing but are not part of all-embracing management change, and that the product or its production process would not be significantly altered to allow Greener marketing. This example demonstrates that companies will have to decide if they aim at compliance, if they want to go beyond compliance, or if they want to base their environmental strategy on win-win solutions, following for example 3M's famous slogan "Pollution Prevention Pays". While Porter and van der Linde (1995) advocate a win-win mind-set in environmental management, Walley and Whitehead (1994) counterargue that win-win solutions should not be the foundation of a company's environmental strategy. They suggest that companies should focus "on the trade-off zone, where environmental benefit is weighed traditionally against value destruction". They argue that safeguarding shareholder value should be the overall objective governing environmental strategies (cf. North, 1996).

Determination of priorities

An experienced industrialist, Winter (1988), proposes four levels of priorities to Green an enterprise.

As first priority, the adoption of whatever environmental protection measures required by law are proposed. As in many countries enforcement of environmental regulations is rather weak, there is a need for self-control by the enterprise. The prevention of acute health hazards should be on the same priority level. Which measures would be appropriate has been explained in section 2.9.

Second-priority measures in Winter's priority list are those which will be of benefit to the company. These could include measures resulting in a net benefit in resource saving which contribute to the creation of new market opportunities or which lead to cheaper processes, to mention just a few possibilities. 3M with its 3P programme learned that the exploration of these second-priority measures, for example, as a result of a company suggestion system, is a real treasure. 3M claims to have saved over US$500 million since the start of its 3P programme in 1975.

Third-priority measures to be included in an environmental action plan concern the adoption of environmental protection measures which will have a neutral effect on the company, i.e. where costs and benefits are fairly balanced. In practice it is not always possible to distinguish third-priority measures from *fourth priority*, which are those measures costing more than they earn (e.g. process changes without legal obligation and without resulting benefit for the company).

A staff commitment is crucial for the success of the action plan. It is advisable to start with actions which lead to short-term success and which have a certain propaganda effect (e.g. energy saving, company cars with catalytic converters, public transport for business travel, environmentally friendly packaging materials, recycled paper in the offices, in-company recycling or selling of wastes for recycling). In the long term only actions which already bear fruit should be started. Winter also argues that environmental protection measures should be introduced according to their consensus potential, in order to overcome possible resistance.

In the following we concentrate on some important steps in defining an environmental strategy and putting it into operation.

Policy statements

There are an increasing number of enterprises that publish their own policy statements or adopt the statements of business organizations concerning environmental protection. The Business Charter for Sustainable Development, explained in the previous chapter, is one such "master" code of practice which could serve as a basis for a company policy statement. Other widely quoted policy statements are the Valdez Principles, which were adopted by a group of powerful investors in the United States following the immense *Exxon Valdez* oil spill in Alaska. Public companies will have to adopt these principles before investors commit their money to a company (Elkington et al., 1991, p. 68).

Policy statements – whether home-made or "brought in" from outside – should cover all aspects of operations, their impact on people, and the role of each part of the company in implementation. The policy statement might read like that of Siemens:

We are committed to conserving the environment and treating national resources with care and respect. This applies as much to our production processes as to our products. The assessment of the environmental effects of our products begins as early as the development stage. It is our aim to go beyond legal requirements and to prevent environmental pollution, or to reduce it to an absolute minimum.

A policy statement does not serve its purpose if it is not followed by concrete actions, which is why the adoption of a policy statement related to the environment or the integration of environmental principles into a company's mission statement should be followed by the production of an environmental strategy.

This formally evaluates the impact of all new products, processes and projects on the physical and social environment so as to minimize such impacts before they occur. The implementation of environmental policy and strategy may require the appointment of an environmental coordinator, sometimes with fairly wide responsibilities and with a direct reporting relationship to senior management, a board member or the chief executive officer.

Commitment counts

Though written statements are an important means of communication, the culture and tone of an enterprise are usually determined by top management and chief executive actions, rather than by any other factor in the organization. The statements they make about the environment will influence the reactions of the rest of the organization and their attendance at senior environmental meetings, and their selection of the agendas will indicate the priorities, concerns and values that are important within the organization.

To raise environmental awareness and commitment General Electric, for example, brought together managing directors who were informed about environmental driving forces and environmental legislation and who then had to cascade the information down in meetings through actions which they themselves designed for their areas of responsibility. As support, the General Electric Management College provided an environmental training pack consisting of a video entitled "Why should I care?", participative tasks, facilitation notes and a glossary of environmental terms. Other companies rely rather on a bottom-up approach to enhance environmental commitment by supporting the individual initiatives of staff.

The omission of health and safety matters from the agenda, or their occasional inclusion, may suggest that they are of little concern to the company. If they are high on the agenda and top management join in the discussions, then these matters are of importance and line managers should act accordingly.

Employees are the ambassadors of the enterprise in the local community just as top management and owners are the ambassadors of the enterprise at government level. Top management will discuss policy and strategic issues with government and appropriate agencies.

In this way, an exchange of views through an effective communication link will ensure that the enterprise is aware of and prepared for changes in environmental regulations which may affect current and future operations and, at the same time, government will appreciate the attitudes and problems faced by industry in complying with legislative changes. In this way, regulatory agencies and management might be persuaded to work together to resolve environmental problems, rather than working against each other.

Top management has a major role to play at a number of levels varying from direct involvement with government environmental policy and controlling the environmental behaviour of managers within the organization, to setting the tone and culture of their environmental concern for the local communities that may be affected. These activities can be supported by statements in annual reports and in providing environmental and social audits along with financial audits. In this way, the annual report becomes more than just a vehicle for financial information, demonstrating the response of the enterprise to its human and social responsibilities. Top management has an important role to play in successfully transmitting the Green image to the outside world. But there are not only large well-established firms which are now spreading the gospel regarding environmental protection; there are also a number of small enterprises which started with a sound environmental commitment and have now grown and become extremely successful due to merging their environmental beliefs with business criteria, not vice versa.

One of the companies which is extremely successful in marketing its Green image is the British company "The Body Shop", which produces and sells naturally based cosmetics and toiletries via shops all over Europe. The Body Shop, led by its charismatic founder, Anita Roddick, is not only a seller of Green products but also a seller of Green ideology. The enterprise has been used, inter alia, as a basis for campaigns in favour of the tropical rain-forest.

A Green business plan

Companies that demonstrate what a "Green" corporate identity really means are still rare in the world of business. The ideas in environmental policy statements must be translated into short-, medium- or long-term business plans. Corporate or strategic planning is concerned with this issue.

Table 5. Environmental SWOT-analysis

1. **What are the environment-related STRENGTHS of the enterprise or business unit?**
   - Environment-friendly products;
   - processes which save resources and do not cause environmental hazards;
   - corporate image as a "Green and clean" producer;
   - staff and management committed to environmental protection;
   - research and development capacities for "clean" products.

2. **What are the environment-related WEAKNESSES?**
   - Products that cannot be recycled;
   - non-recyclable packing materials, bottles, etc.;
   - polluting processes;
   - hazardous wastes;
   - "polluter" image;
   - staff and management not committed to environmental protection.

3. **What are the environment-related OPPORTUNITIES?**
   - Entering new markets;
   - being among the first to offer an "environment-friendly" version of a traditional product;
   - securing long-term survival by shaping a "Green" business image;
   - raising the performance of collaborators by setting a new goal for environmental protection;
   - saving resources (e.g. energy) and costs.

4. **What are the environment-related THREATS?**
   - Environmental regulations require additional investments and could render products unprofitable;
   - increased state intervention in and control of business activities;
   - citizen activist groups take action against enterprises;
   - competitors gain market shares with Green products;
   - identification of staff with the company decreases, retainment and recruitment of personnel becomes more difficult;
   - medium-term survival of enterprise is threatened.

It involves identifying external threats and opportunities, and relating them to the internal strengths and weaknesses of the enterprise. The translation of these results into new plans and directions will enable the enterprise to take advantage of the available opportunities, deal with potential threats, build on strengths and eliminate weaknesses. Environmental criteria should be incorporated into such a self-examination analysis. In table 5 an example of the first step of an environmental SWOT-analysis is given.

A SWOT-analysis is usually a group exercise of high-level management. The purpose of the exercise is to collect, systematize and evaluate the present business situation and to develop scenarios for the future. In general, a number of group sessions are held in which:

strengths,

weaknesses,

opportunities, and

threats

are analysed. Answers for each subject area are grouped according to their importance. For example, in the strength analysis the participants in the group session must reach an agreement on the significance of each listed strength for the further development of the company.

Based on the SWOT-analysis, alternative business scenarios are developed by brainstorming sessions and evaluation. By this process a Green business plan can be shaped.

A business plan including a progressive environmental strategy should lead to all operations in the enterprise being more environmentally compatible and ensure that planning, information and control systems pay equal attention to environmental, economic or financial issues. Environmental impact assessment procedures and information systems should be used for internal and external communication on new development proposals.

Control systems should be as concerned with controlling, measuring and reporting the effects of managerial decisions on the external environment as with the internal financial consequences.

The new Green business plan will require alternative technologies, processes and products to be identified for production as well as for ultimate use and final disposal. The optimum choice combines the economic objectives of the organization, such as least cost production or maximum output with the minimum environmental disruption. Failure to identify and consider alternatives may mean that the preferred least cost option is neither the economic nor the environmental optimum after taking account of the minimum environmental protection measures. The art of creative management lies largely in the ability to identify and select alternatives and to market them adequately.

Measuring environmental performance

Once environmental policies and strategies have been put into operation, the results have to be measured. According to James and Bennett (1994, p. 20), environmental performance is measured to track progress and set appropriate targets, to compare performance internally between sites and/or divisions and externally, to link environmental objectives to business objectives and to monitor the satisfaction of key external stakeholders. Companies would also like to have accurate data on costs, benefits and key environmental parameters of investments. Further, day-to-day activities of plants and departments have to be monitored. Environmental performance is measured not only to provide information for management, but also – or in some cases even more – to satisfy the information needs of stakeholders such as customers, investors, employees, legislators, suppliers, pressure groups, the media, industry associations and the community at large. As the information needs and interests of each of these stakeholders differ widely, a well-targeted communication approach has to be designed (see section 3.5).

What should be measured? James and Bennett (1994, p. 21) state that "the key problem of environment-related performance measurement is converting large amounts of complex data into usable information through appropriate metrics". They distinguish between the following categories of environmental performance measures:

- **Impact measures** track the environmental consequences of business activities on local, regional or global ecological systems. The minimization of such impacts is the ultimate purpose of environmental business management.

- **Risk measures** assess the likelihood and the consequences of environmentally harmful events.

- **Measures of emissions** to air, soil and water are often used as substitutes for impact measures. Because they are required by environmental legislation, these emission and waste measures are common indicators of environmental performance.

- **Resource measures** record consumption of energy, water and other resources.

- **Efficiency measures** relate inputs and outputs and determine, for example, the degree of energy efficiency.

- **Customer-related measures** are concerned with the satisfaction and behaviour of the customers.

- **Financial measures** are concerned with the costs and benefits of environment-related actions of the corporation.

The measurement and review of environmental performance is an integral part of environmental management systems as stipulated by the ISO 14000 series or the European Environmental Management Audit Scheme (EMAS), which will be discussed in section 3.1.

2.2. Eco-marketing: Positioning strategy versus consumer awareness

In many markets, especially the saturated Western economies, the label "environment friendly" has become the new marketing instrument of the late 1980s and 1990s. Those companies that do not take up the environmental challenge are bound to lose their share in the market. Companies who project an environmentally unfriendly or uncaring image will find their business damaged by public disapproval or a consumer boycott.

Companies who mislead the public, or who pretend to be "Greener" than they really are, are not uncommon in the new drive of "Green marketing". This being a relatively new topic, consumers in many countries are aware of environmental problems but they lack information on the environmental impact of the products they buy. Often marketing people and advertising agencies do not know much more than the general public on the subject.

In addition to considering the product itself, marketing which is ecologically oriented has to pay special attention to the distribution channels. This refers to the establishment of efficient recycling systems as well as to environment-friendly transport systems and packaging. Defining a price policy and positioning concept of Green products is another task of the marketing management. Market studies have shown that consumers tend to be prepared to pay higher prices for products which convince them that by using them they are contributing to the protection of the environment. However, these products are not necessarily better than traditional products which do not have a Green label.

If companies do not simply want to continue selling their old products with a Green face-lift, marketing management must define its objectives in terms of the contribution its products and services make to the well-being of the community at large. Specific case-studies of products which have caused harm to the community could be introduced into marketing studies. A far-sighted marketing management will press for research and development to develop new products which are environmentally harmless from cradle to grave.

War of Green arguments

Companies try to win or defend market shares using arguments that are more or less Green.

A French chemicals giant is a case in point. It was beaten to the French market for phosphate-free detergents by a German competitor. The French company, which was Europe's second-largest producer of phosphates, responded by setting out to destroy the Green image of phosphate-free detergents. It plastered thousands of billboards in cities and in the countryside with posters which portrayed dead fish floating in a river – the victims, it maintained, of non-phosphate detergents. But the strategy of attacking its competitor's Green credentials backfired. Many consumers were irritated by the posters and scientists and journalists attacked the French producer's claims as unfounded. Sales of its own products also suffered a consequent slump, according to market analysts.

Worse still, the German non-phosphate detergent came from nowhere to capture 8 per cent of the French market. Another multinational competitor, whose product is Europe's best-selling phosphate-free detergent, has also enjoyed better sales on the French market at the French company's expense.

Source: *International Management*, Autumn 1990, p. 30.

Green consumerism

Consumers are being increasingly supported by Green consumer-product testing. The Council of Economic Priorities of the United States has published a guide, *Shopping for a better world: A quick and easy guide to socially responsible supermarket shopping* (Council of Economic Priorities, 1989). In the United Kingdom the *Green consumers guide*, published annually by Sustain Ability Ltd., has become a best-seller.

Marketing managers should certainly read such guides and journals to anticipate their consumers' reactions. Particularly in retailing, leading supermarket chains increasingly play an ecological gatekeeper function as they set the criteria that producers must meet if they want to be further considered as vendors. Such an environmental "push strategy" therefore supports an environmental "pull" by consumers.

In general, the Green consumer avoids products which are likely to:
— endanger the health of the consumer or of others;
— cause significant damage to the environment during manufacture, use or disposal;
— consume a disproportionate amount of energy during manufacture, use or disposal;
— cause unnecessary waste, either because of overpackaging or because of an unduly short useful life;
— use materials derived from threatened species or from threatened environments;
— involve the unnecessary use of – or cruelty to – animals, whether this be for toxicity testing or for other purposes;
— adversely affect other countries, particularly in the Third World.

Labelling

In many countries consumers identify environment-friendly products by means of environmental labels such as the Blue Angel in European Union countries (figure 3). The labels are often the only criteria the consumer has with which to judge the product.

To unify criteria for labelling, the International Chamber of Commerce (ICC) has suggested a number of principles on how to establish and administer environmental labelling schemes (ICC, 1990; see also RAL, 1990).

Advertising

It is also arguable that advertising, as a medium, is not very effective for dealing with environmental issues. The arguments are often abstract or

Figure 3. Labels for "environmentally friendly" products

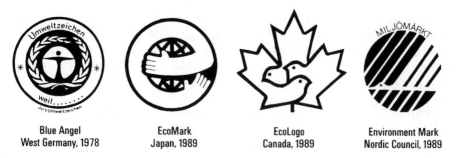

| Blue Angel | EcoMark | EcoLogo | Environment Mark |
| West Germany, 1978 | Japan, 1989 | Canada, 1989 | Nordic Council, 1989 |

bogged down by complex scientific issues. It is arguably easier to seduce someone into buying something with an image of ostentatious opulence than with an intricate explanation of how a disposable nappy manufacturer has eradicated dioxines from the pulp production process. These problems are compounded by the fact that consumers' attitudes to Green products are becoming more complex. Almost all the research into the area shows that, although environmental issues still influence buying decisions, people are much more sceptical about claims made for Green products.

The French *Bureau de la vérification de la publicité* has published a number of objectives and rules (table 6) for the use of ecological arguments in advertising which should be respected by companies who wish to be taken seriously by consumers.

Companies who assume they can afford not to respect these rules could expect an "award". The environmental pressure group "Friends of the Earth" is awarding the "Green Con of the Year" to those companies who are believed to mislead the public.

2.3. Research and development (R&D) management: Environment-friendly products from cradle to grave

At first glance the complex research and development process of pharmaceuticals by a multinational company, and the design of a new chromium-plated restaurant chair by a small enterprise in a developing country have nothing in common. While a new pharmaceutical might reach the market after ten years of research and intensive pharmacological and toxicological testing, the new chromium-plated chair may have been seen in a foreign catalogue by the owner of a small metal furniture manufacturer. The catalogue was given to his foreman to produce a prototype within a week. The metal frame was welded and cleaned using chemical substances, was chromed, and subsequently a leather seat was attached which had been permeated with a spray using chloro-fluorocarbons (CFCs).

Table 6. Objectives and rules for the use of ecological arguments in advertising

The objectives

All reference to the environment must comply with one or more of the following objectives:
- present accurately the significant action(s) undertaken in environmental matters;
- present accurately the positive environmental characteristics of a product;
- provide information on the positive environmental balance of a product;
- provide information in order to modify or correct preconceptions and unfounded or incorrect statements concerning the products, their components and contents.

The rules

1. Publicity should avoid all information that misleads the consumer directly or indirectly on the real advantages or the ecological properties of products, as well as on the actions that the enterprise conducts in favour of the environment.
2. The enterprise must be in a position to produce all evidence to justify its claims, statements or publicity presentations.
3. Publicity cannot resort to demonstrations or scientific conclusions relative to the environment which do not conform to established scientific work.
4. Publicity cannot make improper use of the results of research or of citations taken from technical or scientific works.
5. Publicity must not reproduce or make statements which are not true or linked to the experience of the person making them.
6. Publicity must not give or appear to give a total or complete guarantee of harmlessness in the field of the environment, when the ecological qualities of the product only concern one stage of the product's life cycle or only one of its properties.
7. Advertisements must indicate how the product exhibits the qualities that are attributed to it, and if possible in what context.
8. It must not be claimed that the product presents particular characteristics with regard to regulation and use, if all similar products present the same characteristics relative to the protection of the environment.
9. The claim must not imply a false superiority and/or allow a product to distinguish itself erroneously from other similar products or those which possess similar characteristics in their contribution to the protection of the environment.
10. The advertising enterprise must not take advantage, in an action in favour of the environment, of superiority or anteriority which rests on facts which cannot be objectively verified.
11. A sign or a symbol can only be used in the absence of all confusion on the attributes of a sign, symbol or official label on the subject.
12. The choice of signs or terms used in publicity, as well as the associated colours, must not suggest ecological properties that the product does not possess.
13. If it is impossible, taking into account the difficulties encountered, to justify global statements the publicity will use instead statements such as "contributes to the protection of your environment by ...", "contributes in protecting your environment by ...", "contributes to the environment by ...", and add the necessary details relating to the elements in question.
14. Absolutely no publicity can represent behaviour contrary to the protection of the environment without a positive corrective statement, nor should it incite behaviour contrary to the protection of the environment.

Source: Adapted from *Bureau de vérification de la publicité, 1990.*

Even though the two cases are quite different, they have some common aspects: during the development process, consciously or unconsciously, a number of decisions were made that affected the environment, thus making R&D responsible not only for the technical performance but also for the "environmental performance" of a product.

R&D managers would argue that it is extremely difficult to ascertain today whether a product would be considered environmentally friendly in ten years' time. "We already have enough constraints: costs, quality, deadlines. How could we include this new concern among our R&D tasks?", R&D managers could argue.

It is true that of the products which are on the market today many were not considered to be harmful to the environment when they were designed. On the other hand, R&D staff have already done much – often unconsciously – for the environment. Electric ovens, refrigerators and freezers today need about 30 per cent less energy than 20 years ago. In many machines the power per kilogram of weight has been increased significantly, and more precise machinery allows a more efficient use of materials, to quote just some examples.

In order to systematically consider environmental aspects, an R&D methodology should integrate ecology, economy and technology. An example of such a methodology is shown in figure 4. It contains all the classic process steps starting with the definition of specific R&D tasks and aims and ending with the development of after-sales service. In the following, these steps are discussed in more detail.

The task

R&D may be given the task of reviewing and further developing existing products and processes. This concerns:
□ the design;
□ the applied materials;
□ the manufacturing process; and
□ the product's performance characteristics (e.g. energy consumption).

The Green demand furthermore instigated the development of new products and processes and thus helped to rejuvenate the product range of many enterprises.

Analysis of product life cycle

Once R&D's aims and tasks have been defined, an analysis of the present system follows. This may comprise the existing products of a company or may be extended to the products of competitors. It should start with an

Figure 4. Research and development (R&D) methodology for environment-friendly products

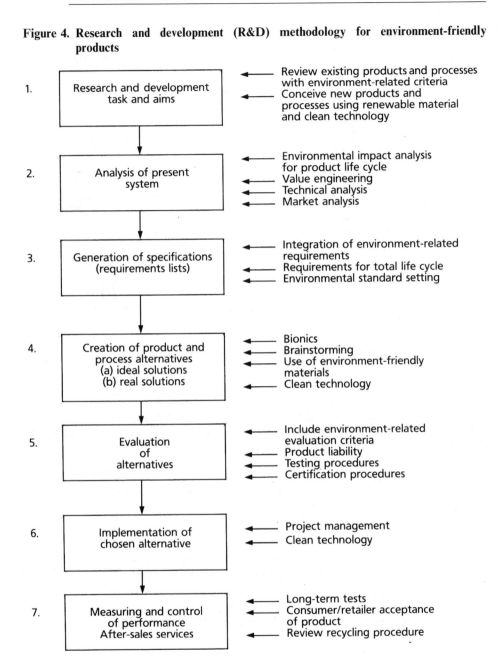

Figure 5. Design of a mixer knife before and after the application of value analysis

Before application of value analysis

After application of
value analysis

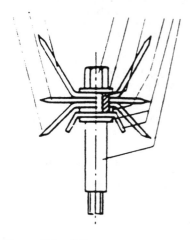

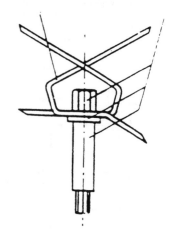

Source: Baier, 1969.

environmental impact analysis of the product's entire life cycle. Table 7 shows such an analysis in a simplified manner. It relates the total cycle of a product's life to the environmental impact caused in each phase. Traditionally, R&D staff have limited themselves to only considering the phase of active use of a product. It is only recently that the whole life cycle of the product has formed part of R&D specifications. The use of renewable materials and "clean technologies" is of importance in all phases of the life cycle. The design of the product should be accompanied by the conception of the recycling procedure.

Besides this *environmental impact analysis of the product's life cycle,* a number of other analysis activities will be carried out. An industrial engineering tool known as *value analysis* or *value engineering* is also particularly suited to making better use of materials and therefore saving not only money but also natural resources. Value engineering analyses the functions or "value" of each element of a product. As a result, products are simplified and reduced to their essential parts. Figure 5 illustrates the effect of a value analysis on the design of a mixer knife.

Specifications for Green products

The third step of our R&D methodology deals with the generation of specifications (requirements list), for both products and processes. In such

Table 7. Life-cycle impact analysis of a product

Life cycle / Impact	Contributes to greenhouse effect	Ozone depletion (use of CFCs)	Acidification (so$_x$ emissions)	Eutrophication	Persistent substances	Nuclear radiation	Use of water / Waste water / Effluent liquids / Pollution of ground water / Impact on sewage / Treatment systems / Cooling water / Discharge	Energy consumption – electric – gas – oil – firewood, etc.	Air pollution, gaseous toxic substances	Landscape destruction / Soil erosion / Forest depreciation	Noise and vibrations / Odour	Dust and particles	Explosions / Spills and leaks / Solid wastes / Hazardous wastes
Design													
Materials handling – raw materials – semi-finished													
Site preparation Construction													
Manufacturing – process I – process II – process III – etc.													
Storage of finished goods													
Distribution (means of transport)													
Sales													
Use													
Maintenance													
Repair													
Storage with client													
Accidents													
Liquidation													
Recycling													

detailed lists the requirements a product or process has to fulfil must be spelled out for each phase of the life cycle. Minimum requirements are those defined by government regulations or the relevant national or international standards (e.g. norms of the International Standards Organization, ISO). Requirements to be given special attention are:

☐ avoidance of scarce, unrenewable materials;
☐ recycling of the product;
☐ energy efficiency of the product;
☐ avoidance of hazardous substances;
☐ minimization of energy, water consumption and polluting substances during the production process;
☐ durability of product; and
☐ product and process quality.

New solutions

After such specifications have been set, the core process of R&D work begins: the creation of new ideas for products and processes. In the first stage – using brainstorming techniques – *ideal solutions* for a given problem should be developed. From these ideal solutions a number of real problem solutions are derived.

In order to develop environment-friendly products, there is much to be learned from nature, which makes the most effective use of its resources. The discipline known as bionics incorporates principles or processes of nature into engineering. Bionics, for example, studies the reasons why trees are so resistant to wind, how birds fly, or how natural membranes function to clean water.

The use of alternative materials and designs is a further potential to be explored by R&D staff. It is at this phase of creation that the manufacture of the products under discussion should be roughly defined and the resulting by-products analysed. Figure 6 shows the case of bicycle manufacture.

Evaluation and testing

Once the product and process alternatives have been developed, they need to be evaluated to see if they really meet the set specifications. Let us assume that a company has a choice between four alternative products, A, B, C or D, and has to decide which product alternative is to be pursued further. To this end, alternatives A to D will be evaluated using "traditional" criteria; costs, performance and design, as well as the "new" environment-related criteria as mentioned in the paragraph on specifications for Green products. The relative importance of each criterion will be weighed and given a ranking

Figure 6. Bicycle manufacturing: Analysis of materials and wastes

	Raw materials and energy	Manufacturing wastes	Treatment wastes
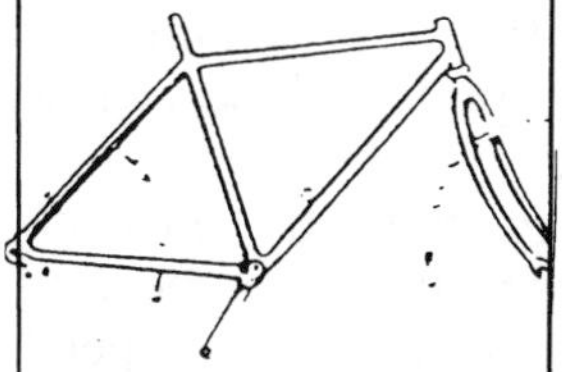	Iron mineral ▼ Steel ▼ Chroming paint	● Slags ● Scoria ● Used baths containing cyanide, chrome ● Paint sludges	Sludges Sludges/Cinders
	Bauxite ▼ Aluminium	Red sludges Fluoride	Sludges
	Petroleum ▼ Plastics ▼ Rubber	Hydrocarbons Catalysers Rejects	Cinders Sludges
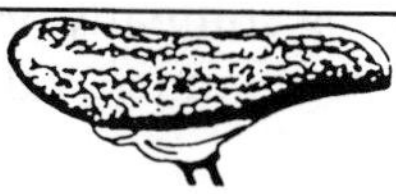	Animal skins ▼ Leather	Organic effluents, acids, oils, chrome	Sludges Cinders
10 Kg		6 Kg	

Source: Centre de documentation sur les déchets, 1989.

according to the alternatives provided. R&D will have to design testing procedures which simulate the whole life cycle of a product.

Implementation

After prototypes or a pre-series have been successfully manufactured and tested in the R&D laboratories or workshops, the responsibility for *implementation* as well as measurement and control of performance will be handed over to Industrial Engineering or Production.

The after-sales services as, for example, the implementation of a recycling procedure, may be commissioned to sales or a specialized engineering department, or even a subcontractor. These arrangements, however, depend largely on size and the organizational structure of the enterprise.

Apart from integrating environmental considerations into each step of the R&D methodology, there are other measures which can assist a company with innovating products and processes faster than their competitors, for example through;

☐ shortening development cycles;

☐ flexible research programmes to keep up with the fast-changing Green agenda;

☐ promoting creativity (e.g. by assigning environmentally committed staff to R&D projects or by incorporating outside researchers or members of environmentalist groups into R&D teams);

☐ integrating an environment-related component into existing company programmes (e.g. quality and productivity improvement programmes, value engineering); and

☐ incorporating environment-related information into R&D information and design systems (for example, into Computer Aided Designs (CAD) systems).

Such a multiple approach will lead to Green products from cradle to grave; however, the readiness to invest and the requirement of cooperation from other departments and management will take time.

2.4. Materials management: Green materials for Green products

In recent years purchasing and materials management emerged as central to achieving a company's production targets and implementing new products and technologies.

Materials management is not only concerned with the planning, purchasing and distribution of materials, but takes responsibility for the whole material flow, starting with the supplier and ending with the recycling of products and materials. Before any material is purchased, material requirements are planned. It is decided if a product or component is to be made or bought and possible suppliers are evaluated. In each of these activities, as shown in figure 7, environmental considerations should have their proper place.

Traditionally, the choice of inputs to an industrial process are based on cost, reliability and continuity of supply, as well as other technical considerations. Environmental management decisions would take account of:

☐ the scarcity of the resource;

☐ the environmental implications of its extraction and use;

☐ the degree of flexibility which the use of substitute resources would offer; and

☐ the ease of recycling.

In practice, resource substitution is often undertaken in response to changes in supply conditions or prices. For example, pressure for the substitution and conversion of hydrocarbons during the 1970s was due to the steep increase in petroleum prices. The "oil crisis" also drew attention to the finite nature of non-renewable fuels, the imprudent exploitation of some natural resources, and the vulnerable position of countries that had to depend upon the import of raw materials.

Environmental check-up for materials

An environment-conscious material planner would like to discover which of the thousands of materials being overseen are harmful to the environment and what substitutes exist. He or she would have to chase up products or processes which contain materials banned by government regulations or are subject to self-limitations in industry. A well-organized enterprise would not only have parts lists for its products or recipes for its processes documented, but would also have materials lists displaying which material is used in what quantities in which products, in order that a materials planner could easily identify the affected products and discuss alternative materials with R&D.

Increasingly, companies establish data banks on materials which contain information on their composition and recyclability.

In order to achieve the right priorities, materials managers are keen to identify the few materials in their inventory which are consumed in high volumes and/or are the most expensive, the so-called "A" products. In this regard, materials management regularly performs Pareto or ABC analysis (see Prokopenko, 1987) of the materials according to cost and volume. Ideally, as a result of such a Pareto or ABC analysis, the materials management team would identify 20 per cent of the total number of materials which represent around 80 per cent of materials costs or volume. They could consequently concentrate their rationalization efforts on these materials.

Figure 7. Environmental considerations in materials management

Materials requirements planning	Make or buy decisions	Supplier evaluations

← Incorporate environmental criteria

Purchasing

Transport of purchased goods to enterprise

← Means of transport

Storage

← Packaging of incoming material
← Leaks and spills

Entry control

← Chemical reaction with environment
← Safe storage
← Clear identification
← Reusable packaging and transport material, e.g. pallets

Material according to environmental specifications and standards

Internal transport

← Means of transport (energy consumption, noise; emissions)
← Packaging, transport devices

Storage

Internal transport

Production

Transport distances →

Number of handlings

Storage of finished goods

Transport of finished goods

Separation of wastes →
Occupational health and safety

Wastes

Storage with retailer

→ → **Waste transport**

See Storage →

Waste storage

Use, storage by client

Disposal

Environmental standards →
Occupational health and safety regulations

Waste treatment, recycling

Recycling

Why couldn't materials managers also carry out an "environmental Pareto analysis"? The new form of Pareto analysis would rank materials according to their environment friendliness or harmfulness.

Though most companies would not be able to classify all of their products strictly according to these criteria, they should be able to group them into broad categories. The most environmentally harmful materials group would certainly contain toxic substances, which pose threats to the health and safety of the producers as well as the users. For example, such materials contribute to the greenhouse effect (e.g. CO_2), ozone depletion (CFCs), acidification (SO_x), and so on, and would also fall into this category. The least harmful or most environmentally friendly materials would be those constituting no health risk, and which are easily recycled.

After an environmental Pareto analysis has identified the 20 per cent of the materials range which is most harmful to the environment, materials management would make it a matter of priority to find substitute materials or alternative processes.

Dealing with suppliers

A simple, but not environmentally friendly, way would consist of subcontracting the activities which involve substances harmful to the environment. Guidelines for the "make or buy" decisions of a company should therefore stress that environmentally harmful materials, components and processes should be eliminated rather than subcontracted or given to a subsidiary in a developing country which has less strict environmental regulations, and a less conscientious population. Many companies today carefully evaluate their suppliers concerning quality, production organization, skill of personnel and technology. In such evaluations environmental criteria are also increasingly included.

In a survey on the Greening of purchasing (North, 1995), an overwhelming majority of the German manufacturing firms interviewed considered environmental criteria as very important or important for purchasing decisions. Their motives to do so ranked as follows: legal requirements, image considerations, ethical motives. Costs ranked only fourth.

When a contract between supplier and client is signed, the mode of delivery forms part of the agreement. At this point a client could insist on environment-friendly packaging and means of transport.

Materials handling and storage

Materials will be stored for varying periods of time before an entry control is performed or at different stages of the production process.

Stores should be safe from a health and safety point of view, the material should bear clear identification, and means of transport (e.g. boxes, pallets) should be standardized and reusable. Chemical reaction of the material with the environment (rust is a very common one) should be avoided, as well as leaks and spillages. The respective training of transport and warehouse workers is indispensable. Special precautions have to be taken in the case of hazardous materials. Information on how to do this can be found in a useful guide entitled *Storage of hazardous materials*, issued by the UNEP Industry and Environment Office.

Materials management, often called logistics, holds the responsibility for the transport of materials in many companies. Transport distances depending on the more or less favourable layouts of plants, number of handlings of a material in successive production stages, means of transport (energy consumption, noise and exhaust emissions, electric fork-lift versus diesel fork-lift) and type of transport containers play an important role in environment-friendly materials management.

A relatively new task for materials management is the handling of all kinds of wastes and not just the traditional selling of shred in the metal industry or food wastes in the food processing industry. Waste management and recycling will be dealt with in more detail in section 3.7.

2.5. Production management: Inputs, throughputs, outputs

Production management has the task of transforming inputs with given specifications and resources into the desired output. However, desired outputs are often associated, as previously mentioned, with undesired outputs such as pollution and waste. In the transformation process itself, employees and the surrounding community may be affected. The objective of achieving high overall productivity, in terms of a high output/input ratio, requires making the best use of resources, including environmentally relevant factors such as material and energy. Hence production managers are directly concerned with environmental issues in their operational responsibilities.

Looking into the main tasks of production management, we can trace how environmental concerns are affected. In production management we may distinguish between two sets of tasks, firstly, those dealing with setting up and continuously improving the production system and, secondly, the day-to-day production planning and control tasks. Figure 8 displays the first category of tasks and gives examples of environmental considerations to be taken into account. In Figure 11 (page 70) the second set of "day-to-day" production management tasks are displayed.

Figure 8. Environment-oriented production management: General tasks

Definition of production philosophy	Investment planning	Implementation of new equipment, technology and work organization	Continuous improvement of systems performance
New philosophies in line with environmental protection	Clean technology, and resource saving as investment cause	Workers' and management training Commitment to environmental protection Environmental impact assessment	Good housekeeping
Simplification	Environment friendliness as decision criteria	Testing of new equipment according to environmental criteria	Resource saving programmes
Inventory reduction			Suggestion systems
JIT	Cost-benefit of environment-related investments	Decentralization of environmental responsibilities to supervisors and work groups	Performance review of production system
Product vs process orientation	Environment-related production costs		

Eco-efficiency as production philosophy

In discussions, the environmental aspects of production management are often associated with leaks and spills or clean technology. The decisive rule, however, of a production philosophy is often forgotten.

The current production philosophies aim at a simplification of products and processes, inventory reduction and just-in-time (JIT) manufacturing. A process orientation also contributes to a better use of resources. Eco-efficiency, cleaner production and lean production are based on a common philosophy: to reduce "waste" in all steps of a production process. Eliminating waste will lead to improvements in eco-efficiency and thus contributes to:

☐ less energy consumption;

☐ less wasted material;

☐ less materials handling; and

☐ less intermediate storage.

Investing in cleaner production

In investment plans, short-, medium- and long-term purchases of production equipment and the siting of production facilities are laid down. Environmental criteria should be incorporated, therefore, when these investment plans are established. Investments should not only be implemented to replace old equipment with new, faster and more precise or automated

Figure 9. An imaginative cost-benefit optimization for alternative investments to reduce the level of pollution

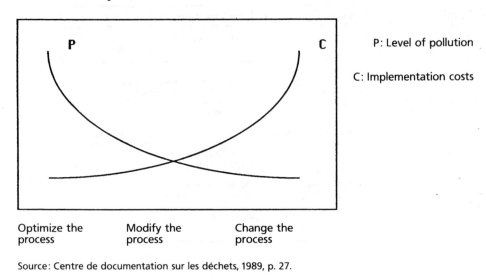

P: Level of pollution

C: Implementation costs

Optimize the process   Modify the process   Change the process

Source: Centre de documentation sur les déchets, 1989, p. 27.

processes. Resource saving, especially energy saving and cleaner production, is one of the major investment causes of the 1990s.

Which investment decisions will be taken – optimization, modification or a complete change in production process – largely depends on the calculated or perceived costs and benefits (see section 2.7). Figure 9 shows an example of a cost-benefit curve for different investment alternatives to reduce the level of pollution:

☐ optimize existing process;

☐ introduce process modifications; and

☐ change process completely.

Figure 10 displays an example of pollution and waste reduction by improved process control. The payback period of this innovation implemented in a British cement plant in the late 1980s was only three months.

Before investment decisions are taken, an environmental impact assessment should be applied to identify and evaluate impacts arising from the planned production process, and possible mitigating measures such as the adoption of a less hazardous storage system, revised layout, or process modifications.

The implementation of an optimized, modified or completely changed production process may be associated with a number of additional adverse environmental impacts due to the fact that:

☐ the process is operated under sub-optimal conditions;

Figure 10. Cement kiln pollution and waste reduction by improved process control

Background

The manufacture of cement in its present form was patented in 1824. Known as Portland cement, it requires the burning of fuel together with limestone and clay, yielding a clinker which is then ground with gypsum to give cement.

Burning is carried out in a rotating, inclined kiln. The process is a complex one, in terms of the reaction chemistry, the thermal conditions in the kiln and the dynamics of the process. The firing temperature largely determines the quality of the product cement. However, both the NO_x and SO_x levels increase with higher temperatures.

The process must, therefore, be operated within a certain band of temperature with the optimum at the lower end. If the process is operated too far below this optimum, an unusable product is generated. If the temperature is too high the coal is wasted, cement quality reduced and air pollution increased.

There are many possible disturbances to the process – for example, changes in the calorific value of the coal and the combustion of the feed, which make it difficult to operate manually.

Clean technology

The LINKman expert-system monitors continuously all the appropriate process variables such as the flue gas temperature, oxygen, NO_x level and the power used to turn the kiln. It then makes adjustments to the coal, air and feed rates on the basis of a model of the plant's behaviour derived from operational experience. The system can also make smaller adjustments, more frequently. This allows the plant to be run much closer to its optimum conditions than is possible under manual control. One significant novel feature of the instrumentation is the measurement of the NO_x level in the flue gas, which gives valuable information on the temperature in the firing zone.

Enabling technology

The system has been made possible by improvements in:
– the science of expert system control; and
– measurement technology, which has led to a reliable and sensitive NO_x analyser.

Advantages

– The wastage of coal at high temperatures is avoided.
– Higher quality clinker is produced.
– The clinker requires less energy to grind.
– The lining of the kilns has a longer life.
– NO_x and SO_x emissions are reduced (a NO_x level of around 500 ppm is typically reduced to 200 ppm).

Economic benefits

These figures relate to a LINKman installation on two kilns at a British cement plant.

Cost saving	£/year
Coal	500,000
Grinding clinker	430,000
Total	**930,000**
Capital investment	**£203,000**
Payback	**3 months**

Figure 10. *(cont.)*

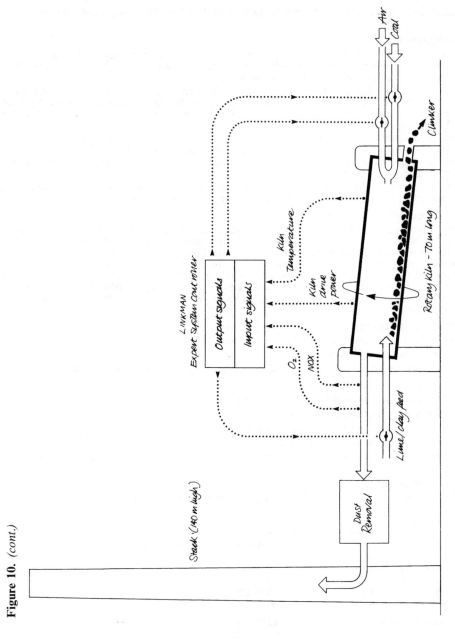

Source: Reproduced with the kind permission of the Controller of Her Britannic Majesty's Stationery Office.

☐ management and workers lack experience with the new equipment and facilities; and

☐ a monitoring programme is not yet functioning.

At this stage of starting up production management, managers should be particularly careful that the standard set for the process is respected and that systems performance is continuously improved while meeting the daily production targets.

Production planning and control

In production planning and control, which might be seen as production management proper, environmental considerations have to be incorporated into planning, scheduling, production preparation and control, quality planning and control as well as into auxiliary services such as maintenance and recovery of resources (see figure 11). Allocate the right equipment for the right task, reduce set-up operations, avoid reheating or restarting of processes under sub-optimal conditions, and avoid accumulation of environmental hazards: these are all guidelines for the establishment of the yearly, monthly, weekly and daily production planning and scheduling.

Production control is not only responsible for the achievement of the desired outputs in terms of quantity and quality, but also has to control working practices, resources consumption, emissions, and the flow of (hazardous) products and wastes. Process variables have to be carefully controlled, particularly in process industries, not only for the sake of the end product but also to avoid undesirable emissions which could easily lead to a withdrawal, by government authorities, of permission to run a process.

Monitoring and improving systems performance

The formulation and implementation of a monitoring programme will provide data on potential adverse environmental effects, and act as an early warning system for approaching unacceptable pollution levels and the need to review mitigation measures. Monitoring may take many forms, from continuous air and water quality measurements to random biological tests of sensitive species and effects on neighbouring communities. The programme may be independently operated by the company, a regulatory authority, an independent organization, or a combination of these. In all cases, the plant production management should keep a close watch on the monitoring results as a means of maintaining optimum efficiency of the facility with minimum impact.

Environmental audits, which are described in detail in section 3.3, are a particularly useful tool to keep track of the environmental effects of production.

Figure 11. Environment-oriented production management: Planning and control

Planning of production programme	Scheduling, production preparation	Production control	Quality planning and control	Auxiliary services, e.g. maintenance, recovery of resources
Eliminate environmentally harmful products	Right equipment for the right task (criteria: resource saving, pollution prevention)	Respect and process rules and "good working" practices	Consider product and process quality	Consider emissions and leaks, energy consumption, noise, etc.
Consider seasonal variations of environment (e.g. climate)	Avoid accumulation of environmental hazards	Control resources consumption and recovery of resources	Identify environmental consequences of process or product malfunction	
Reduce set-ups reheating, energy losses		Control emissions		
		Control flow of hazardous products and wastes		

Following regular audits or monitoring reports, there are many ways to improve systems performance. Some of these are described below.

Maintenance

Environmental performance can be improved by preventive maintenance of equipment and facilities, leading to improved operating efficiency, conservation of resources, and reduced incidence of plant failure. Similarly, maintenance is a prerequisite in preventing catastrophic environmental effects from a major accident in a hazardous facility.

Improved maintenance:

☐ reduces pollution of air and water (e.g. less emission if optimal process is attained; revision of pipes, tanks, filters and catalysts functions well only when process parameters are kept within small tolerances);

☐ reduces energy consumption (e.g. clean tubes, adequate cooling substances, clean heat exchangers, greasing of running gear, isolation);

☐ reduces the generation of wastes, in particular toxic wastes (e.g. by ensuring uninterrupted processes, optimal process temperature, functions of filter, catalysts);

☐ uses raw materials more productively (e.g. avoids evaporation, material losses, reduced scrap);

☐ reduces dust and noise (e.g. by cleaning, greasing, replacement of damping material, cleaning of filters and factory floors); and

☐ extends the life span of equipment (e.g. by running equipment according to specification, adequate maintenance strategy).

Total quality management (TQM)

The need to take care of process and product quality in an integrated manner is now widely accepted. This corresponds to the total quality management (TQM) approach. TQM is a useful tool for pollution prevention and resource saving, as many companies have demonstrated. AT&T is a case in point. There, total quality environmental management (TQEM) is practised in order to improve environmental performance not only in manufacturing but also in offices. Other companies are also raising environmental issues in their quality circles or asking their quality specialists to identify possible environmental consequences of process or product malfunction. TQM will not function without a committed workforce. We shall discuss the quality approach to environmental management in section 3.1.

Workers' participation and commitment

The challenge to minimize spills and leaks by good housekeeping, for example, requires the assurance that workers are aware of the consequences of pollution, and the potential dangers to the health and well-being of their families. Workers should be encouraged to suggest improvements both to the working environment and for reducing the level of pollution emanating from the plant. Such suggestions should be accompanied by some reward and encouragement scheme.

Supervisors are well suited to taking up environmental responsibilities in production. They directly oversee operations and can influence their team to act in an environmentally responsible manner. They are often the first to recognize leaks, spills or malfunctions of equipment. Given their role with the production organization, supervisors are in a good position to promote environmental protection on the shop floor, provided they receive adequate training.

Cooperation

Production managers should cooperate with suppliers and in-house engineers to improve the technical efficiency and cleanliness of existing and new production equipment. Industrial engineers and ergonomists could be particularly helpful in improving worker/equipment interaction. Production managers should also work closely with materials management to find ways of selling or disposing of residual materials, such as solid wastes and used solvents, rather than simply dumping them.

Production managers and their staff could reduce potential problems and conflict with the community by becoming more involved in community projects, sitting in on community committees, making company meeting facilities available, and in the use of equipment and tools for community projects. Such links are of mutual benefit to the company and the community.

2.6. Project management: Taking care of environmental concerns

With increasing regularity we are witnessing public announcements for new projects, be it a new plant, a testing circuit for cars, a hotel complex or a garbage incineration plant. These public statements by company managers typically react to rumours already spread by press and television. A public debate is opened while the enterprise continues to pursue its plans and usually starts construction or engineering works before a final conclusion has been reached. Due to public opposition, or intervention by

Figure 12. Tasks of environmental project management in the different project phases

Project phases	Integrate environmental protection into project	Win acceptance for project	Develop project according to specifications, time, budget	Establish and monitor performance standards
Extension	///////////////	///////////////	X	///////////////
Operation	X	X	X	///////////////
Implementation construction	X	X	///////////////	X
Conception/ planning	///////////////	///////////////		Establish performance standards

Project management tasks ⟶

Key:
X = Project management tasks to be accomplished in respective project phase.
//// = Indicates most important tasks for each phase.

environmental or other public agencies, a project may undergo modifications and additional installations may be requested, for example, to prevent undue emissions. Budgets are overspent and project delivery deadlines pass unmet.

This scenario occurs throughout the world. However, those involved on a day-to-day basis in the enterprise are the project managers and engineers whose tasks are to deliver projects on time, within budget and to specification.

How can they cope with the environmental challenge?

To successfully complete a project, environmental considerations have to be included in all project phases such as:

☐ conception and planning;

☐ construction and implementation;

☐ operation; and

☐ extension.

There are four main tasks of environmental project management which need to be carried out in each of the project phases, as shown in figure 12:

☐ integrate environmental protection into project;
☐ win acceptance for project;
☐ develop project according to specifications, time and budget; and
☐ establish and monitor performance standards.

Let us now examine these environmental project management tasks individually.

Integrate environmental protection into project

The optimum solution of economic benefits to the company at minimum environmental cost to the community is most likely to be reached when environmental issues are incorporated into project planning at the earliest possible stage. During the conceptual phase of project design radical alternatives may be considered to assist imaginative solutions to possible environmental problems. At this stage, environmental impacts will not be defined or evaluated, but consultations with experts may indicate the nature of potential environmental problems. It is better to rethink the project with a view to avoiding potential problems than to undertake a detailed redesign at a later stage.

During the early design stage, projects should be screened to determine the need for an environmental impact assessment (EIA, see section 3.2), which can be expensive and time-consuming, but no more so than the late discovery of serious environmental impacts requiring a redesign of the project to minimize them. It is also wasteful of resources if minor projects are subjected to environmental assessments which deal only with trivial issues. The screening process, which can be undertaken in consultation with the relevant public authorities, should identify the need or otherwise for an EIA.

For such a screening, the "activity impact matrix for project management" (see figure 13, for an example) could be used. The matrix is, however, applicable in all phases of the project. The procedure for assessing the potential impact of projects on the environment starts with an analysis of the environmental setting of the project. This requires a baseline inventory of the physical components and natural processes likely to be affected and how they, in turn, might affect the project. For example, a water inventory would detail the hydrological systems, erosion, siltation, temperature inversions, eutrophication, flora and fauna breeding grounds, and so on.

An analysis of the social environment might include the dependence of a community on its physical and cultural resources such as food, raw materials, beliefs and lifestyle. By making site visits, the project managers, accompanied by small groups of appropriate experts, will be able to appreciate the potential problems both for the environment and the project.

Win acceptance for the project

This is of crucial importance at the *very early* stage of the project, before mistrust and the NIMBY phenomenon arises (NIMBY = Not In My Backyard).

The management task of informing those who may be affected should be approached as a positive project planning activity rather than an interruption or interference. First of all, the management and staff of a company have to be convinced about the utility of a project. But this is not enough. To be successful a project must rely on the support of the local community, who generally supply the labour force and other services. This is illustrated by the following example of sensitive public relations involved in the development of a low-grade gold mine in a developing country.

The developer introduced three public participation programmes to monitor changes taking place in the lives of the local population: to provide information on the project and its implications; to obtain the views of the people on the changes that were taking place; and to assist them in maximizing the project's benefits and minimizing its costs.

The public relations group maintained regular contact with villages. They handled compensation claims and payments, conducted information patrols with film, video, written and spoken material, and coordinated requests for engineering services, such as road access, water supplies, power and building materials, to the villages.

The environment group monitored the project's impact on the physical environment and extended this to the subsistence resources of the region and the health and nutritional status of the population.

The local business development group guided the formation of village businesses by providing essential services to the project and its community. Businesses established included fruit and vegetable supply, transport services, vehicle maintenance, hotel accommodation and earth-moving contracting.

These and other actions taken to increase the community's understanding of the benefits of the project contributed to the overall success of the development.

The aspect of communication in the context of environmental management will be treated in more detail in section 3.5 of this book.

Develop projects according to specifications, deliver on time within budget

As projects usually tend to take longer and cost more than foreseen, the construction and implementation phase is, also from the environmental point of view, extremely critical. As time runs short and costs rise, project workers, contractors and subcontractors are pressed to proceed faster and

Figure 13. Activity impact matrix for project management

Characteristics of the existing environment	Nature and characteristics of the proposed development
	A. Site preparation and construction — Industrial buildings and process structures / Highways, roads and tracks / Bridges / Railways and sidings / Transmission lines, pipelines and corridors / Barriers, including fences / Channel dredging and straightening / Canals / Dams and impoundments / Deepwater ports and marine terminals / Blasting and drilling / Underground works / Surface excavation, including cut and fill / Land clearing, including burning / Surfacing or paving / Erosion control and terracing / Landscaping / Noise and vibration
1. Climate and air quality — Wind directions and speeds / Precipitation/humidity / Air quality	
2. Water — Hydrological balance / Groundwater regime / Drainage/channel pattern / Sedimentation / Flooding / Water quality / Surface waters	
3. Geology — Unique/special features / Tectonic/seismic activity and volcanic activity / Mineral resources / Physical/chemical weathering / Landslide / Subsidence	
4. Soils — Erosion (wind and water) / Slope stability / Liquefaction / Bearing capacity / Settlement/heave / Earthworks / Soil structure	
5. Ecology — Species checklists / Plant communities / Diversity (species and spatial) / Productivity / Biogeochemical/nutrient cycling	
6. Environmentally sensitive areas — Prime agricultural land / Forestry land (silviculture) / Wetlands/estuarine and coastal zones / Landfills (solid/toxic waste disposal sites)	
7. Land use and land capability — Land use / Land capability	
8. Noise and vibration — Noise / Vibration	
9. Visual quality	
10. Archaeological, historic and cultural elements — Archaeological structures and sites / Historic/cultural structures, sites and areas	

Figure 13. *(cont.)*

Nature and characteristics of the proposed development						
B. Process operations Cooling water discharge; Liquid effluent—industrial and domestic; Water demand; Septic tanks—domestic and industrial; Stack and exhaust emissions; Spent lubricants; Noise and vibration; Odour	*C. Raw material handling* Dust; Stockpiling; Noise and vibration	*D. Energy-producing operations* Atmospheric emissions; Cooling water discharge; Liquid effluent; Water demand; Stockpiling of materials; Noise and vibration	*E. Transportation requirements* Highways, roads and tracks; Bridges; Railways and sidings; Shipping	*F. Accidents/hazards* Explosions; Spills and leaks; Operational failures	*G. Waste disposal and control* Landfill, spoil and overburden; Solid waste disposal	

Source: Adapted from UNEP/IEO, 1980, p. 50.

often with less care, or to use cheaper materials which may later cause environmental damage, such as oil leaking into the ground.

The implementation of projects involves a number of project management issues: large, negative and unforeseen delays and redesign problems may occur because of tighter administrative rules or community protests. Here, early contacts with the administration are advisable to avoid repeating a very common practice of first creating situations and then negotiating.

Establish and monitor performance standards

Once a project has become operational, performance standards must be monitored. The value of a daily or weekly production report is well understood and accepted. Why should an enterprise not have a weekly environmental report where the amount of liquid and gaseous emissions, the tons of wastes and the energy consumed are related to production output or to the environmental standards set by government regulations or internal company regulations (see section 2.5 on production management)?

This would not only create awareness within the company and motivate company staff, but also prove the credibility of an environmental policy to the general public.

Project management in the siting of new industrial plants

The siting of new industrial plants very often causes controversies related to environmental issues, and conflicts tend to persist during all phases of a project.

A decision on the siting of any new, large industrial plants must take account of the carrying capacities of the supporting ecosystems and of socio-economic structures and resources. Decisions on siting production facilities must also take account of the long-term costs and benefits involved.

Industrial site selection depends upon a wide range of factors, such as minimizing transport costs for materials and finished products, conveniently located power, water and other supplies, proximity to the client, available workforce, etc. Other sites may be preferable to the community in terms of their capacity to assimilate the environmental and social impacts of particular types of project.

In the environmental assessment of industrial sites four steps may be identified as follows:

☐ prepare a short list of potential sites which offers choices in relation to the objectives of both the developer and the community;

☐ identify and compare the sites in terms of a common set of criteria which express the degree to which change can be absorbed without excessive environmental degradation;

☐ relate the assimilative capacity of each site to the planned mitigation measures at the plant; and

☐ compare the socio-economic impacts between sites. In some cases, public or private expenditure on physical and social infrastructures may offset undesirable effects. Some socio-economic effects may be amenable to modification.

The availability of strategic land-use plans and designated industrial development zones can be of assistance in the selection of suitable areas for industrial development. Permits for developments within designated zones can usually be obtained more quickly than those which are not designated. Consequently, developers and their advisers should be aware of land-use planning constraints in those areas in which they propose to site a new facility. For any given area, there may be a number of published land-use plans which set out agreed policies to safeguard areas.

The reader may have formed the impression that this section on project management is only relevant to big industrial projects such as the establishment of new plants. While environmental project management is of special importance in such cases, it is equally valuable for minor new developments such as the installation of a modified production line or the redesign of painting facilities in furniture manufacturing.

2.7. Financial management: Make pollution prevention pay

The financial management of a company, including financial and accounting rules, can have a major impact on corporate performance in terms of growth and survival and of pollution and environmental impact. Environmental management can and does produce financial benefits, but these may only occur in the longer term. For example, savings from recycling waste products may not appear on the balance sheet for a number of years. The pressure of meeting quarterly profit targets can jeopardize future growth and the ultimate survival of an enterprise. It can also jeopardize the future prospects of the enterprise in taking proper account of its environmental responsibilities. In summary, prevention is better than cure and early investment in clean technology can avoid later environmental problems and enhance profitable performance.

The pitfalls of environmental economics

A financial manager could agree with this last statement but would certainly like to calculate the degree to which pollution prevention pays. That may not be so difficult in the case of investing in a new furnace which consumes 20 per cent less energy than the existing one. It may be rather more difficult, however, in the case of a new product that completely changes the product mix.

In addition, there always remains the question of which portion of the "real" costs of resources consumption or pollution is borne by the enterprise and which portion is paid by the present community, or by future generations. While markets seem to function reasonably well in pricing scarce resources, the costing of environmental damage is still very much in its infancy. To make the polluter pay will require an answer to the question of the time cost of environmental damage (see Cairncross, 1991; Pearce et al., 1989). Attaching a price to environmental values can admittedly be difficult. However, where this can be done or approximated, charges, taxes, pricing mechanisms and other economic measures can play an important role in internalizing such costs (UNCED, 1991). As long as enterprises are not charged for polluting or pay only minimal prices for their resources, they will have no financial motivation to change. However, environmental awareness in most countries is rising and with it the costs of natural resource consumption as well as charges for polluters and generators of waste (see section 1.3). Financial managers should therefore carefully observe the debate on environmental economics aimed at defining values and prices for using the environment.

Investment calculation

Owing to the dynamics of a business, calculating whether pollution prevention pays is not an easy task. After the strategic reorientation of an enterprise to Greener products, it is normally impossible to calculate what the results would have been if the change had not taken place. Better sales figures could be attributed to the new strategy without an unequivocal cause-effect relation.

As environmental regulations are undergoing dynamic development in many countries, an investment today will have to rely on a number of assumptions concerning, for example, pollution charges and environmental taxes or emission rights which will affect cost-benefit calculations. Financing costs (interest rates, credit lines) are another factor which determines whether pollution prevention pays. As a positive development, some banks provide more attractive financing conditions for Green investments compared with others. Legal provisions may also grant shorter depreciation periods for environment-related investments.

Remembering that in private enterprise the objective of financial management is to maximize returns on capital invested, a short-term view of finance puts particular emphasis on the validity of such tools as discounted cash flow. Discounted cash flow has many valid applications, but, used non-selectively, it will always recommend minimizing expenditure in the short term, even though greater expenditure might then be necessary in the longer term. The long-term expenses will be discounted away, while the short-term expenditures will be predominant. There will always be an argument against spending money now to save money in the future on water, energy, raw materials and pollution control. The choice of discount rate is fundamental to calculations of present value of investment, and it is rare for the corporate discount rate to be the same as the social discount rate. While current and short-term returns are valued highly, society may have a longer-term perspective, valuing future returns more highly than corporations. The core of calculating returns on environmental investments is reconciling conflicting discount rates.

All the above-mentioned factors have to be taken into account if we want to calculate accurately if pollution prevention pays. But this should not be an excuse not to invest in the environment.

There are many examples of the ways pollution abatement can be achieved practically without incurring costs. A French study has shown that in 70 per cent of the cases studied, changing the process technology to prevent pollution eliminated the pollution and resulted in lower operating costs than those of the original polluting plants.

Furthermore, the investment strategy is of importance. For example, after the initial installation of pollution control equipment, the marginal cost of pollution abatement is small. Alternatively, the implementation of pollution control, through good housekeeping measures, may result in low initial marginal costs, but in a steep rise in abatement costs, when material or process design changes are required, or with the installation of new equipment. An earlier replacement of machines and installations could also result in economic and ecological benefits.

While in one case an earlier replacement of equipment will be ecologically and economically feasible, in other cases equipment with a longer life expectancy may be the best bargain. If a decision is taken to buy one machine that will last ten years instead of buying two machines during the same period which will last five years each, the resources required to build one machine, as well as to shred or recycle it, will certainly have been saved. Such considerations are feasible only for equipment and facilities where the technology involved does not advance too rapidly.

Table 8 illustrates such an example. The overall costs of two machines with different purchase prices and different life expectancies (duration of service) are evaluated by a static cost comparison (linear depreciation). The comparison results in lower overall costs of machine 2, which costs 20 per cent more than machine 1 but has a 25 per cent longer life expectancy.

Another possibility, in evaluating whether pollution prevention pays, is calculating the pay-back period of a new Green process in relation to the old one. Figures 14 and 15 outline a simplified method of performing such an evaluation. Which pay-back period is considered satisfactory depends on the life expectancy of the equipment and facilities, as well as financial regulation and the velocity of change in a specific industrial sector. While in the process industry a pay-back period of three years may be satisfactory, equipment in the electronics industry may have to pay off in five months.

Controlling

Financial controlling is perhaps the most effective function in monitoring and enforcing environmental targets in the enterprise in terms of input-output or cost-benefit indicators. The task of environmental controlling is to base its actions on a set of indicators relevant to the environment. The selection of meaningful indicators for environmental performance is a particular challenge. Some of these indicators are displayed in the box on page 84.

Table 8. Cost comparison between two alternative machines

	Machine 1	Machine 2
I. Data for cost comparison		
(1) Cost of acquisition (DM)	100 000	120 000
(2) Duration of service (years)	8	10
(3) Reselling price (DM)	10 000	12 000
(4) Produced units per year	15 000	15 000
(5) Calculatory interest rate (%)	7	7
(6) Other fixed costs (DM/year)	1 000	1 100
(7) Labour costs (DM/year)	6 000	4 800
(8) Material costs (DM/year)	1 500	1 500
(9) Variable manufacturing costs (DM/year)	1 850	1 850
II. Cost comparison		
(10) Linear depreciation	$\dfrac{100\,000 - 10\,000 =}{8}$ $11\,250$	$\dfrac{120\,000 - 12\,000 =}{10}$ $10\,800$
(11) Interest on bound capital (average)	$\dfrac{100\,000 + 10\,000 \times 0.07}{2}$ $= 3\,850$	$\dfrac{120\,000 + 12\,000 \times 0.07}{2}$ $= 4\,620$
(12) Sum of fixed costs (DM/year) (10 + 11 + 6)	16 100	16 520
(13) Sum of variable costs (DM/year) (7 + 8 + 9)	9 350	7 900
(14) Total costs (DM/year) (12 + 13)	25 450	24 420

III. Result: Machine 2 will be purchased owing to the lower total cost for the number of units to be produced (machine utilization) as fixed in (4) above.

Source: Blohm et al. 1982, p. 140.

Figure 14. Calculation of pay-back of investments: Part I

Investments excluding taxes (Year:)	Former process (Cost)	New process (Cost)
Total investment (excluding property and equipment subsidies)		
Investments	1:	2:

Annual operating cost, excluding taxes, of the production line concerned	Former process (Cost)	New process (Cost)
Variable expenditures – raw materials – energy – water – manpower – manufacturer – maintenance – miscellaneous expenditures – pollution royalties (water and air agencies) Total variable expenditures *Fixed costs* – factory overheads – local taxes – insurance Total fixed costs		
Operating costs	OC1:	OC2:

Earnings, excluding taxes of the production line concerned	E1:	E2:

Investment recovery (pay-back time)

Source: Centre de documentation sur les déchets, 1989, p. 50.

Figure 15. Calculation of pay-back of investments: Part II

In order to compare various projects, two types of criteria are likely to be used:	$I_2 =$	Investment needed to set up the clean process.
☐ those which are based on discounting; and	$I1$:	Investment needed to continue production using existing process or a new, but standard, process.
☐ those which avoid having to use discounting techniques; investment recovery time, presented below, is an example of this type of comparison.	$OC2$:	Annual operating cost of the clean process.
	$OC1$:	Annual operating cost of the former process.
Investment recovery time (pay-back time) is defined by:	$E2$:	Annual earnings with the clean process.
	$E1$:	Annual earnings with the former process.

$$\frac{I_2 - I1}{(E2-OC2) - (E1-OC1)}$$

Source: Centre de documentation sur les déchets, 1989, p. 51.

Some indicators to be set by environmental controlling

$$\frac{\text{units produced}}{\text{energy consumed}}$$	$$\frac{\text{production losses due to environmental problems}}{\text{period of time}}$$
$$\frac{\text{units produced}}{\text{waste produced}}$$	$$\frac{\text{planned pay-back of Green process}}{\text{real pay-back of Green process}}$$
$$\frac{\text{units produced}}{\text{materials consumed}}$$	$$\frac{\text{planned costs of Green process}}{\text{real costs of Green process}}$$
$$\frac{\text{units produced}}{\text{water consumed}}$$	

As the implementation of environmental policy in an enterprise is a gradual process, in subsequent planning periods "tougher" environmental indicators could be set as targets and monitored by controlling.

The Greening of accounting

With tightening environmental regulations the roles of accounting and reporting are likely to change. Norway, for example, already requires a statement in the annual report of the board of directors on the impact which the enterprise has on the environment and the measures which it has taken in this area. Even though legislation has made enterprises specifically liable for environmental damage and reparation in many countries, very few companies have recognized such liabilities in conventional accounting terms. Financial statements currently only provide a very partial assessment of the full monetary and environmental operating costs and give little insight into the state of the natural resources upon which companies are economically dependent. (EIASM, 1991).

The Greening of accounting involves a very different consideration of environmental expenses, liabilities and resource costing (Koos, 1991).

Operating expenses related to environmental protection are generally recorded immediately but may, in some countries, be deferred over a reasonable period. Significant environment expenses may, in some countries, be considered as an extraordinary item in the Profit and Loss Statement.

Capital expenses related to environmental measures may, in most countries, be depreciated on an accelerated basis. Estimated removal and site restoration costs are generally charged to income over the same period as the related property, plant and equipment are amortized. There exists a generalized trend to consider expenditures for environmental measures as immediately deductible for tax purposes. In several countries tax credits are available for enterprises implementing certain methods of energy conservation.

Information disclosure for *environmental liabilities* and accruals for environmental matters will become more and more common in most countries. Whether the expense related to these liabilities is deductible for tax purposes varies from country to country.

There are important limits to the assignment of a monetary value (price or cost) to certain resources, and in no country are the externalities (such as pollution and waste arising from the exploitation and use of certain resources) taken into account by the Profit and Loss Statement.

A lot of research is currently being undertaken on the ways to evaluate resources in a market economy and the solution, rendered mandatory in future years, might well consist of obliging the enterprises to use imposed mandatory values to account for externalities.

In the training of financial managers and accountants, emphasis should be placed on the environmental and other implications of short-term oriented decisions and on examining the potential areas of application of costing projects throughout the life of the project on a non-discounted basis. More attention should also be given to those situations where the purpose of investment is for growth rather than for short-term profit. In short, there is a need to re-examine the basic proposition that the alternative to a given project is to put the money in the bank and for the company to go out of business as a manufacturing and marketing enterprise and become an investor.

Financial and accounting managers should be made aware of the external, community and economic effects of pollution and environmental degradation, and should consider the implications for national economic growth as well as for the growth of their enterprise and the specific financial implications of waste.

Data on environmental economics in general, and the internalization of environmental costs, together with information on damage costs, cost-benefit analysis and the result of cost-benefit studies, will show how the total benefits of a clean environment outweigh the costs of achieving that clean environment. A primary aim would be to establish a closer working relationship between financial managers and accountants, on the one hand, and project and environmental engineers within the corporation, on the other.

2.8. Personnel management: Developing human resources for the environment

A recruitment officer is interviewing a candidate for a vacancy and has discovered that, though highly suitable with respect to many of the requirements, the candidate is active in the Green movement. Would you hire such a person? The candidate, after having been questioned, inquires whether the enterprise undertakes any action to protect the environment. Will you be able to give a true and credible answer?

Personnel management has a central role to play in implementing the environmental policy of the enterprise. And it is personnel who will receive negative feedback if the enterprise only pays lip service to environmental protection. How then could personnel management integrate environmental considerations into its functions? This is discussed below and summarized in figure 16.

Recruitment

Environmental thinking starts at the recruitment stage. Many enterprises now use their Green image to advertise vacancies. When interviewing candidates or evaluating them in assessment centres, environment-related questions could be included. In the initial interview the recruitment officer should state that the enterprise is committed to environmental protection, if it is the case, and explain what the expected behaviour of the new employee will be. The first briefing, after the employee begins work, should also address questions regarding environmental responsibilities and relevant contact persons in the enterprise.

Retaining staff

Retaining qualified staff is becoming more and more dependent on the image of the company. No one would like to work for a company that has a polluter image. On the contrary, the Green image of a company can contribute to retaining personnel. However, there are many other factors contributing to the retention where environmental considerations could be included: training, remuneration and incentive schemes, work organization, works council and trade union relations, social services and personnel administration; all of them will be discussed below.

Career development

Career development, the third classic function of personnel management, must demonstrate in a tangible way that ecological behaviour pays off. Environmental training and the activities of a staff member could be a good argument for promotion. Community work concerning environmental questions should be rewarded as well as the credibility and ecological consistency of middle and upper management, which comes directly to the question of corporate identity, staff motivation and awareness. Personnel must ensure that all company members are aware of company mission statements or policies concerning the environment. To create and maintain a certain level of awareness, the enterprise journal may include a periodical environment section. The families of staff members as well as the community in general could be involved in environment-related activities.

Tools of personnel management

Training is undoubtedly a key factor in developing human resources for the environment. There are three separate and distinct elements to environmental training and education. The first is a need for knowledge of the environment, of actions taken and of the consequences of the actions for the quality of the environment. The second concerns the attitude towards environmental issues which holds a key to appropriate environmental behaviour. The major challenge for personnel managers lies in providing environmental education which will bring about a change in attitude and behaviour among managers, staff and workers. The final purpose of environmental training is the acquisition of the relevant skills; general environmental management skills as well as special skills (see table 9). Training issues will be discussed in more detail in section 3.6 of this book.

Remuneration and incentive schemes may be detrimental to the environment either implicitly or explicitly if incentive schemes are only output related, and if they do not take into consideration the resources consumed

Figure 16. Green personnel management

Recruitment

☐ when interviewing candidates or in assessment centres, include environmental questions;

☐ state that your enterprise is committed to environmental protection and explain expected behaviour of candidates;

☐ advertise vacancies using a Green image.

Retention

☐ promote a positive ecological image as a motivator to retain personnel;

☐ ensure retention by ecological incentives.

Career development

☐ demonstrate that ecological behaviour pays off;

☐ enforce compulsory environmental training courses before promotion;

☐ ensure credibility and ecological consistency of superiors;

☐ reward community activities.

to produce this output. Hence the reason why enterprises develop environment-related incentive schemes which consider materials and energy savings. Some of the traditional job evaluation systems result in higher evaluations if the workers are exposed to noise, vibrations, radiation, toxic substances and other adverse environmental factors. In these cases, instead of paying compensation for negative environmental factors, working conditions should be improved.

Suggestion schemes or other ways of participation constitute a further incentive to contribute to the improved environmental performance of an enterprise (see section 3.5).

Work organization and working time regulations can also contribute to the protection of the environment. Flexitime in larger companies, for instance, contributes to the reduction of traffic congestion and facilitates the use of public transport. Decentralized environmental responsibilities raise awareness and permit faster action. In some sectors, especially the process industries, the continuous running of machines and installations reduces energy consumption and waste. In these cases, personnel management must collaborate with the works council to develop suitable shift systems which guarantee the smooth running of the process as well as taking into account the interests of the workforce.

Any environmental policy of the enterprise can only be implemented efficiently if the workforce, by means of their representatives, collaborates with management. In some enterprises the management and works council or union have set up a joint environment committee. The environment issue should be part of a collective agreement, for example, in the case of applying new job evaluation methods or the installation of an environment-related

Table 9. Personnel management tools to Green the enterprise

Corporate identity — staff motivation and awareness	Training	Remuneration and incentive schemes	Work organization and working time	Works council and trade union relations	Social services	Personnel administration
Company mission statements to all	Environmental education	Ecological incentives	Flexitime	Joint environment-related activities	Organize car sharing	Mode of business travel
Enterprise journal with environment section	Functional training programmes	Revise job evaluation and re-muneration systems	Decentralize environmental responsibility		Promote use of public transport	Performance appraisal system
Involve families of staff members	Integration of environment into trainee programmes and profes-sional training		Continuous work	Include environment in collective agreements	Healthy canteen food	Promotion credits
Community work				Assist works council to acquire environmental competence and profile	Environment-related books and journals in library	
					Participate in community activities	

incentive scheme. It is not necessary for personnel management to always take the initiative in environmental issues; it should, however, assist the works council to acquire an environmental profile. These issues will be discussed in more detail in section 4.2.

Within the *social services* of a company there is wide scope for environmental activities. Car pools to travel to work could be organized, and special arrangements for public transport could be negotiated. Many companies already offer health foods in their canteens and stock environment-related books in their libraries.

Groups of staff members are encouraged to take part in community activities; for example, equipment may be provided to clean public areas or plant trees. The enterprise grounds themselves could be Greened with planted tress and bushes, and so on.

Environmental commitment and responsibilities at 3M company

"Working in our favour" is 3M's official environmental policy. The goals expressed therein provide a foundation for goals set at the operating level. Also, success has been achieved by assigning specific responsibilities for the programme, by making people accountable, and by recognizing their success. The Vice President of Environmental Engineering and Pollution Control (EE&PC) is accountable to the chief executive officer and the board of directors; individual environmental engineers are accountable to the head of EE&PC; and technical personnel are held accountable by voluntary group pollution-prevention goals. Groups and individuals alike are recognized when they meet established goals or succeed on individual (Pollution Prevention Pays) Projects.

Finally, our employees' own environmental concerns, the general increase in the public's environmental awareness, and other environmental programmes at 3M are all working in our favour.

These factors increase employee interest in "daily awareness of" opportunities for pollution prevention. Every day, more employees are recycling paper, glass and cans at home. At work, too, they hear regularly about 3M's recycling and energy conservation efforts. Hundreds of 3M commuters save energy every day, riding to and from work in our 105-van Commute-A-Van fleet. And, with two wastebaskets in every 3M office – one for high-quality recycable paper – every person at 3M must decide – almost by the minute – whether to recycle or not. Through repeat exposure, 3Mers are learning to keep the idea of pollution prevention in mind.

Source: Bringer and Benforado, 1989, p. 125

Personnel administration could reinforce environmental policy by laying down specific administrative rules. In Switzerland, for example, many companies provide their employees with a half-price rail card to be used for business travel but which is also valid for private travel.

Initiatives of this kind are becoming more widespread.

2.9. Occupational safety and health: Managing the working environment

The protection of workers from diseases and injuries arising out of their employment is a major management responsibility which has to be pursued by coherent efforts to improve both the working environment and the general environment.

A growing concern in occupational safety and health (OSH) is the increasing number of chemicals used in all types of manufacturing. Acute and chronic intoxications, allergies or cancer after long-term exposure to certain substances reflect only some of the consequences of adverse environments on the workforce and population. But there are also the more traditional threats: in some industrial sectors up to 10 per cent of the workforce show signs of deafness following noise exposure. Industrial accidents may have negative consequences not only for the workforce but also for the population, as Bhopal, Seveso and Chernobyl have all demonstrated. A number of major industrial accidents have been attributed to human error, i.e. the inability of the respective operators of complex systems to act appropriately.

What can management do to protect workers and the general population from hazards associated with industrial activities?

ILO Conventions on the working environment

Convention Concerning the Protection of Workers Against Ionizing Radiations, 1960 (No. 115).

Convention Concerning Protection Against Hazards of Poisoning arising from Benzene, 1971 (No. 136).

Convention Concerning Prevention and Control of Occupational Hazards Caused by Carcinogenic Substances and Agents, 1974 (No. 139).

Convention Concerning Protection of Workers Against Occupational Hazards in the Working Environment Due to Air Pollution, Noise and Vibration, 1977 (No. 148).

Convention Concerning Occupational Safety and Health and the Working Environment, 1981 (No. 155).

Convention Concerning Safety in the Use of Asbestos, 1986 (No. 162).

Convention on Safety in the Use of Chemicals at Work, 1990 (No. 170).

Compliance with OSH standards

While compliance with current occupational health regulations is essential, many companies set higher internal standards which incorporate a safety factor more stringent than the authorized daily exposure to a particular substance. Such a policy demonstrates concern for the employees' welfare as well as providing a safety margin against exceptional exposures. Compliance with standards requires close cooperation with labour inspection as well as with OSH committees within the company.

Prevention of health hazards

Instead of repairing unfavourable working conditions it is far more effective to prevent hazards by carefully examining, at the design phase of systems, the relationship between worker and environment. This is important not only regarding the impact of the working environment on the worker, but also regarding the effect of human actions on the environment. The layout of control panels or the design of process control software to match human control abilities and capacities has become an important discipline, and the study of human error contributes to this. Human errors which subsequently may lead to an environmental catastrophe, or even only to a small leak, are due to a failure to detect a relevant signal (e.g. on a control panel), incorrect identification and interpretation, and incorrect action selection. Tracing back human error and critical incidents is one of the standard procedures of occupational safety and health.

Management should be particularly sensitive in investigating minor critical incidents which, combined with adverse factors such as specific climatic conditions and operator fatigue, could lead to major industrial accidents.

Screen workforce regularly

Regular medical screening of the workforce, and of those workers with long-term exposure and high-dose exposure to specific environmental hazards, is a prerequisite following preventive measures. Special groups such as young workers, women, older workers, rehabilitees and persons with disabilities should be screened at shorter intervals or should undergo specific medical examinations.

Monitor exposure levels

It is not enough only to screen workers. Their medical status report must be associated with the working conditions to which they are exposed. Such a measurement programme would be conducted to estimate individual workplace exposure to hazardous substances, and to assist in the assessment of the efficiency of engineering and process controls (Rossiter and El Batawi, 1987). In this context the control of exposure limits is of prime importance. Exposure limits, the definition and legal status of which vary from country to country, refer to concentrations or intensities at the workplace which in repeated long-term exposure, even up to an entire working life, do not in general lead to health impairment of either the workers or their offspring. The exposure time is usually calculated as eight hours per day and 40 hours per

---

Ergonomic systems design

Ergonomics is a discipline concerned with creating optimal working conditions. Its principles should be applied to create a working environment which does not harm workers. This can be illustrated by the following example.

An operator in the chemical industry controls a process by means of a control panel. The work object is a chemical substance which is synthesized during the process. To protect the operator from heat, noise and hazardous substances an air-conditioned cabin has been designed where the control panel is located. In this case, the operator is not directly exposed to the hazards of the process while remaining in the cabin.

The operator, however, may influence the working environment of colleagues and of the population, depending on his/her control skills of the process. It is obvious that the presentation of information on the control panel, especially the feedback of corrective activities, is of immense importance for avoiding errors and thus preventing process deviations which could lead to undue pollution or accidents. Ergonomic systems design deals with the optimal design of such workplaces.

---

week (for more detailed information see the ILO *Encyclopaedia of occupational health and safety*, 4th ed., in preparation). In cases where actual workplace concentrations exceed exposure limits, remedial action has to be taken.

Eliminate or reduce the exposure of workers to health hazards

To achieve this there are basically four principles which may be applied. The first priority is to control, eliminate, substitute or reduce *at the source*. If a substance cannot be eliminated by changing the process, a less harmful replacement should be sought. In the case of noise exposure, the duration of operation or the amount of machinery working at one time could be reduced.

If it is not possible to eliminate, substitute or reduce, the *health hazard should be isolated* to avoid or restrict its propagation and transmission. Ventilation intercepts and captures the emissions before they have a chance to enter the workplace. Ventilation systems should be equipped with filters or other effective treatment facilities to prevent the hazardous substances, even in low doses, polluting the outside environment.

Environmental factors affecting the worker

☐ Thermal environment

The factors which control the heat exchange between the body and the environment determine whether a particular thermal environment is hazardous. Temperature, moisture content (relative humidity), and velocity of the surrounding air, and type of clothing worn are all important. Age, weight, physical fitness, degree of acclimatization and workload are additional factors. Physical work contributes to the total heat stress of the job.

☐ Noise and vibrations

Noise generated by equipment, processes or work practices within the workplace will affect people both within the workplace and outside it. Noise levels likely to be experienced by persons working in industry are calculated in terms of an eight-hour noise exposure. It has been found that exposure should not exceed 85 dBA if long-term hearing damage is to be avoided.

In addition to noise, the related phenomenon of vibration has also been found to affect occupational health, depending on frequency, intensity and point of application to the body.

☐ Radiation

The effects of electromagnetic radiation on the body depend on wavelength, duration and power of the radiation exposure. Radio transistors, microwaves used in radar communications and certain types of cooking, and diathermy equipment are all capable of causing significant heating of tissues.

Visible radiation is obviously important in the workplace: good lighting conditions are essential both for working and to avoid accidents.

Ultraviolet (UV) radiation is associated with industrial processes such as electric arc welding and can cause eye inflamation ("arc-eye") if proper eye protection is not provided.

Lasers emit beams of coherent radiation of a single wavelength and frequency which is highly collimated and hence has a large energy density in a narrow beam. The eye is vulnerable to injury from laser light and hence direct viewing of the laser should be avoided.

Nuclear radiation is the most hazardous type of radiation, as exposure to lower doses shows only long-term consequences. In many countries there exist detailed regulations for the health protection of persons exposed to nuclear radiation.

☐ Chemical agents

Metals: Many countries have stringent legislation to control exposure of workers to heavy metals such as lead, mercury and cadmium, and acute poisoning from these causes is now largely confined to accident situations. However, concern continues to be felt about the possible chronic effects of these materials.

Solvents: Many such compounds have been identified as important factors in the chemical exposure of workers. Since most are volatile, and many also penetrate the skin, exposure is easily produced wherever they are used. Apart from the narcotic effect, specific solvents have been found to cause toxic effects.

Pesticides: Pesticides and their health effects both for human and animal species in the natural environment have been in the forefront of public attention ever since the days when DDT residues were found to be adversely affecting the breeding of raptorial birds in Europe and North Africa. Since that time the nature of the compounds used in agriculture and public health has changed to less toxic substances.

Source: Based on Rossiter and El Batawi, 1987.

If the isolation of the health hazard is not enough to reduce workplace concentration below exposure levels, or if it is not feasible, there remain two principles which could be applied.

Working persons should be protected or isolated from the process. Protective gear, air-conditioned cabins and even remote control operation are possible means to avoid contact with an adverse environment.

Be prepared for emergencies

Emergency preparedness forms part of the OSH strategy. Because of its importance for the environmental performance of an enterprise, section 3.9 is devoted to this issue.

The use of chemicals at work

A particularly critical area of OSH is the use of chemicals at work.

This includes the production, handling, storage, transport, disposal and treatment of chemicals, and the release of chemicals resulting from work activities, as well as the maintenance, repair, and cleaning of equipment and containers for chemicals.

In 1990 the International Labour Conference adopted a Convention (No. 170) and a Recommendation (No. 177) concerning safety in the use of chemicals at work. To help put these instruments into practice, a code of practice, *Safety in the use of chemicals at work*, was published by the ILO in 1993. Conventions, Recommendations and codes of practice are useful tools for management to evaluate and improve their present use of chemicals:

Chemicals should be classified according to the type and degree of their intrinsic health and physical hazards.

All chemicals should be marked so as to indicate their identity. Hazardous chemicals should in addition be labelled, in a way easily understandable

---

to the workers. The information to be given on the label should include, as appropriate: trade names; identity of the chemical; name, address and telephone number of the supplier; hazard symbols; nature of the special risks associated with the use of the chemical; safety precautions; identification of the batch; a statement if a chemical safety data sheet giving additional information is available from the employer; the classification, hazards, safety precautions and emergency procedures.

Further, a number of *operational control measures* for the safe use of chemicals should be taken. These include all the measures which have been cited above in this section. The box below, displaying some protective measures, is an extract from the ILO code of practice which contains more detailed information on operational control possibilities. Further information on this subject is given in section 3.9, "Prevention of industrial disasters".

Control measures to provide protection for workers

(*a*) *Good design and installation practice*:

(i) totally enclosed process and handling systems;

(ii) segregation of the hazardous process from the operators or from other processes;

(iii) plants, processes or work systems which minimize generation of, or suppress or contain, hazardous dust, fumes, etc., and which limit the area of contamination in the event of spills and leaks;

(iv) partial enclosure, with local exhaust ventilation;

(v) local exhaust ventilation;

(vi) sufficient general ventilation.

(*b*) *Work systems and practices*:

(i) reduction of numbers of workers exposed and exclusion of non-essential access;

(ii) reduction in the period of exposure of workers;

(iii) regular cleaning of contaminated walls, surfaces, etc.;

(iv) use and proper maintenance of engineering control measures;

(v) provision of means for safe storage and disposal of chemicals hazardous to health;

(*c*) *Personal protection*:

(i) where the above measures do not suffice, suitable personal protective equipment should be provided until such time as the risk is eliminated or minimized to a level that would not pose a threat to health;

(ii) prohibition of eating, chewing, drinking and smoking in contaminated areas;

(iii) provision of adequate facilities for washing, changing and storage of clothing, including arrangements for laundering contaminated clothing;

(iv) use of signs and notices;

(v) adequate arrangements in the event of an emergency.

A selected occupational safety and health bibliography

For managers interested in more details of OSH the following ILO publications are recommended (for details see the Bibliography):

- *Encyclopaedia of occupational health and safety* (*Fourth edition*).
- *Higher productivity and a better place to work* (*Trainers' manual*).
- *Radiation protection of workers* (*ionising radiations*).
- *Protection of workers against noise and vibrations in the working environment.*
- *Occupational exposure to airborne substances harmful to health.*
- *Safety in the use of asbestos.*
- *Prevention of major industrial accidents.*
- *Major hazard control.*
- *Safety and health in the use of agrochemicals: A guide.*
- *Safety in the use of chemicals at work.*
- *Safety, health and welfare on construction sites.*
- *Ergonomic checkpoints.*
- *Recording and notification of occupational accidents and diseases.*

2.10. Managerial advice from outside: The environmental consultant

Have you ever thought of inviting a Green pressure group, perceived by many industrialists as a threat, to advise your enterprise on environmental policy? Though many managers would rather avoid getting involved with Green activities there are many examples of environmental pressure groups advising company management. As the result of an accident a Swiss chemical company had greatly polluted the Rhine, which in turn had killed many fish. The enterprise decided to collaborate with an international environmental pressure group and seek its advice on environmental questions. This may not be the most common case for environmental consulting, but it does reflect a change in the attitude of management towards the handling of environmental issues.

Why should a company hire environmental consultants and what do they offer?

☐ Management may feel uneasy about how to handle environmental questions in the enterprise, where to begin and how to define company policy and proceed systematically.

☐ Management may seek specialized advice or intensive professional help on a temporary basis, as it is often difficult to hire environmental specialists.

☐ In internal discussions on the pros and cons of certain new Green products or processes, consultants would provide an impartial outside point of view.

☐ Consultants are brought in to support interest groups in the company.

☐ Consultants moderate processes of change or induce learning processes in the organization.

☐ They assist in advertising the Green image of a company, or help to raise awareness within the company.

☐ Consultants evaluate environmental risk and costs associated with a planned acquisition or merger of companies. Environmental liabilities and costs have become major criteria for the assessment of a financial investment.

This is by no means an exhaustive list of the reasons why companies use the services of consultants.

Environmental consultancy is a growing business. Assistance to repair, audit and certify companies wishing to implement environmental management systems, as stipulated by British Standard 7750, ISO 14000 or the European Environmental Management Audit Scheme, is provided by a wide range of consultancy firms. Established management consultancy firms have added the environment to their existing range of services, while a number of new consultancy firms specialized in the environment have been created.

Advertising agencies are busy in "Greening" product images. Audit houses have added environmental audits to their portfolios. Law firms assist companies in preventing liabilities and in lawsuits following environmental damage. Investment consultants analyse the environmental performance of companies, which then becomes a major investment argument. Insurance companies have created specialized units to assess the environmental risks associated with insurance contracts and provide advice to their clients on how to reduce environmental risks. Engineering firms specialized in water purification, waste treatment, environmental monitoring, recycling, retrofitting, and energy saving, to name just some examples, are booming. Some companies have set up consultancy services to sell the know-how gained in cleaning up their own company to other firms.

Environmentally minded enterprises have set up joint consultancy circles to assist them in environmental questions and to learn from one other. In Germany, for example, some companies have formed the Environmental Management Association BAUM. Within BAUM a number of working parties, staffed by specialists from the participating enterprises, exchange experiences and jointly develop solutions for environmental problems.

The International Chamber of Commerce (ICC) provides advice, mainly concerning clean technologies, to enterprises in developing countries.

As we have already mentioned, some environmental pressure groups are providing advice to companies on environmental matters (see Chapter 4). Environmental consultants not only work for companies but also for environmental protection agencies and local authorities charged with the enforcement of environmental regulations. This is why companies are increasingly facing competent counterparts who are in a position to assist in implementing environmental policies.

These numerous factors result in a heterogeneous mix of environmental advice which companies may have to decide on. The box below displays Ten Commandments for clients, summarizing critical points of which a potential user of consultancy services should be aware.

The Ten Commandments of using consultants

1. Learn about consulting and consultants!
2. Define your problem!
3. Define your purpose!
4. Choose your consultant!
5. Develop a joint programme!
6. Cooperate actively with your consultant!
7. Involve the consultant in implementation!
8. Monitor progress!
9. Evaluate the results and the consultant!
10. Beware of dependence on consultants!

Source: Kubr, 1996, p. 721.

Table 10. The three generations of organizational environmental management

	First generation	Second generation	Third generation
Management & strategic context	• No long-term strategic approach • EM seen as an overhead cost • Pollution control oriented	• Limited strategic approach • EM seen as a necessary survival cost, especially to avoid future liabilities • Regulatory oriented	• Essential element of a company's strategic approach • EM seen as a fully legitimate business expense • Competitive advantage, quality, community and customer oriented
Philosophy	• Reactive – existing legal requirements only • Minimum expenditure possible	• Reactive, some strategic consideration • Holistic approach to expenditure at plant level	• Proactive, active participation in the regulatory development process • Holistic approach to expenditure at company level, with some costs hard to differentiate from normal business activity
Organization	• Plant manager driven in response to regulators • Buy in expertise	• Director and plant manager driven in response to regulators • Establishment of in-house expertise, appointment of an environmental manager	• Dynamic, multiple interactions and drivers • Extensive and widely disseminated in-house expertise
Strategy	• No link to business strategy	• Considered as a liability in business strategy • Generally case-by-case approach	• A key integrated aspect of business strategy, with positive and negative considerations
Tools	• EIA (if a statutory requirement)	• Auditing	• LCA, EMS • Databases • CBA and environmental factors, TCA
Operating principles	• Respond to threats as they emerge	• Consider at all stages of operation • Build relationship with regulators	• Fully integrated at all stages • Highly developed internal and external communication
Funding	• As forced	• As forced, with additional funding if economic conditions dictate	• As necessary to meet business objectives with established funding priorities
Resource allocation	• Locally set	• Locally set with some central assistance	• Variable throughout organization based on a corporate philosophy appropriate to the business area
Targeting	• End of pipe	• Full process oriented	• Full process and LCA oriented
Priority setting	• As forced by regulators	• Strategic regulatory analysis	• Competitive advantage and business planning analysis
Measuring results	• Filed and forgotten • No QA	• Incorporated in company and plant operational performance data • Some QA	• Data based • QA and readily available to all
Evaluating progress	• Plant level	• Formal plant and company reporting system • Board level • Formalized reviews	• Integrated as part of the company's overall performance • Regularly reviewed and assessed for completeness

EM = environmental management; EIA = environmental impact assessment; QA = quality assurance; LCA = life cycle analysis; EMS = environmental management system; TCA = total cost assessment; CBA = cost-benefit analysis.

Source: A. Krol: "Environment management – Issues and approaches for an organisation", in M. D. Rogers (ed.): *Business and the environment* (London, Macmillan, 1995), pp. 51–90. Reproduced with kind permission.

3

How to make your business lean, Green and clean

This chapter deals with:

☐ the implementation of environmental management systems (EMS);

☐ the most important management tools to achieve this objective; and

☐ specific areas of action such as waste management, energy saving and emergency preparedness.

3.1. Environmental management systems (EMS)

While the previous chapter dealt with the integration of environmental considerations into all different management areas, this chapter will provide management tools which are of relevance to most of those areas. In this respect the content of this book is structured as an analogy to the matrix organization of a company.

We are witnessing important changes in the approaches to environmental management. These changes can be described as three generations of organizational environmental management, as shown in table 10, Krol, 1995). The first generation of environmental management was reactive, concentrating on existing legal requirements only; there was no long-term strategic approach. Environmental management was seen as an additional overhead cost and was oriented mainly towards pollution control. Companies responded to threats as they emerged. The second generation of environmental management is characterized by a limited strategic approach, where environmental management is seen as a necessary survival cost especially to avoid future liabilities, and is still regulation oriented. Leading corporations worldwide are now attempting to implement a third-generation approach, where environmental management is an essential element integrated into a company's strategy. Environmental management is seen as a fully legitimate business expense. Corporations have a quality-, community- and customer-oriented focus regarding environmental issues. This requires active participation in the regulatory development process and a holistic approach to environmental expenditures at company level. It is this third-generation approach which requires the building of institutional environmental management capabilities based on coherent and comprehensive environmental management systems (EMS). EMS have become a means of operationalizing self-regulation which companies increasingly prefer as compared to government-imposed regulations.

Environment: A new quality dimension

To integrate environmental considerations into mainstream business activities, the total quality approach seems to provide an adequate tool. This growing interaction between corporate environmental management and quality management is most advanced in the United States, where leading companies have established the Global Environmental Management Initiative (GEMI) which has created the TQEM (total quality environmental management) approach. Why is TQEM relevant to environmental management? James (1996) argues that the stress placed on the importance of customers provides a useful framework for considering and responding to the demands of environmental stakeholders. TQEM's emphasis on commitment to continuous improvement is helpful to organizations wishing to move beyond mere compliance with environmental regulations. Its focus on eliminating the root causes of problems rather than the symptoms fits with the growing awareness of pollution. Prevention is often the best approach to environmental problems. TQEM's belief that quality is everyone's responsibility within a company fits well with the growing awareness that all employees have to make a contribution to environmental performance.

Companies wishing to build an EMS on quality foundations can choose between different options. In 1992, the British Standards Institute published its BS7750 standard for environmental management systems, which was based on the existing BS5750 standard for quality management systems. Since 1996, the ISO 14000 series which closely resembles BS7750 has been approved. There is now an international standard on EMS available that translates the quality philosophy into an environmental context. Just as ISO 9000 has become essential for many companies, accreditation to the ISO 14000 standard will be required for companies around the world who want to export, import or tender for international contracts. The European Environmental Management Audit Scheme (EMAS), which applies to sites rather than to entire organizations, is in its philosophy and procedures closely related to the above-mentioned EMS. While the EMAS is favoured by regulators and environmental administration, the ISO standards are more prominent in business interactions.

Environmental management systems have a number of structural elements in common:

- **a definition and description of EMS:** including organization structure, tasks, auditing and accreditation procedures;

- **an environmental policy:** a definition of the mission, policy and targets of environmental management;

- **environmental impacts:** measurement and evaluation of business activities' effects on the environment;

- **communications and reporting:** a description of internal and external communication processes;

- **an environmental programme:** objectives and measures to implement the environmental policy, including deadlines;

- **operational control:** a description of a system to operationalize and control the implementation of the environmental programme;

- **environmental management records:** documentation of environmental effects, measures taken and results;

- **internal audits:** periodical review of systems and their performance;

- **education and training** to ensure that staff understand the aims and tasks of an EMS and are competent in performing their tasks.

The fulfilment of the above criteria as stipulated by ISO 14000, BS7750 or the EMAS may lead to an accreditation of an enterprise or site by external auditors. Even though internationally standardized EMS may lead worldwide to comparable and coherent systems, environmental performance, however, might still differ significantly. In this respect James (1996) remarks that only a minority of TQM schemes are delivering their anticipated benefits, because organizations do not understand that successful TQM is not a set of tools and techniques but a radical management philosophy. Introducing management systems without an underlying commitment to change organizational values is unlikely to succeed. A more conceptual criticism of TQM is that its focus on continuous improvement can block a recognition that sustainability requires radical, discontinuous change.

Tools for action

To implement an EMS a number of management tools are needed. The Environmental Challenge Scan introduced in Chapter 1 can serve as a checklist to determine which management area is a priority area to introduce change. The environmental SWOT analysis assists in defining policies and strategies. The present chapter will introduce the tools that will put the action plan into practice:

☐ An environmental impact assessment before new operations start.

☐ An environmental audit by which the environmental performance of the company and its units can be assessed, weak points can be identified and subsequently the required mitigating measures may be determined.

☐ An appropriate organization of environmental functions in the enterprise ensuring that initiatives are taken, expertise is available and responsibilities are clear.

☐ Communication and participation inside and outside the enterprise, which play a decisive role in the creation of a positive company image to avoid resistance and commit staff to contribute to a common cause.

☐ Environment-related training which raises awareness, introduces behaviour changes and helps to acquire the necessary managerial and technical skills.

These tools for action will be dealt with in the following sections.

3.2. Environmental impact assessment (EIA)

The EIA is a management tool to:

☐ forecast the impact that a project will have on the environment; and

☐ find ways to reduce unacceptable impacts.

In principle, an EIA should apply to all actions likely to have a significant environmental effect. The potential scope of a comprehensive EIA system is, therefore, considerable and could include the appraisal of policies, plans, programmes and projects. Policies need not, in themselves, be environmental but they may have environmental consequences.

Application

At the project development stage, the available options are often severely limited by earlier decisions made at a higher level. Misspecification of a project assessment may occur if the higher-level policies were not subject to such evaluation.

Assessment of individual projects can only be conducted.once proposals have been made. It cannot guarantee optimum site selection, and a thorough assessment of all alternative actions may be prohibitively expensive and time-consuming.

The scope of viable alternatives decreases at the project level; the willingness to contemplate alternatives may also decline.

The available time for the collection and analysis of environmental data will become increasingly restricted at the lower tier unless a programme of establishing environmental baseline data is undertaken independently of individual project EIAs.

When projects are individually small in size, but collectively large in number, an EIA at the plan and programme stage may lead to a reduction in the time required for evaluation.

Reasons for carrying out environmental impact assessments (EIA)

☐ Assurance of adequate procedures for managing environmental risks, and compliance with procedures.

☐ Improved statutory compliance.

☐ Identification of environmental risks and problem areas, early warning and prevention of potential adverse environmental effects (risk identification, assessment and management).

☐ Improved financial planning, through the identification of future and potential capital, operating and maintenance costs, associated with environmental activities.

☐ Improved preparation for emergency and crisis situation management.

☐ Improved corporate image and positive public relations.

☐ Enhancement of environmental awareness and responsibility throughout the corporate hierarchy.

☐ Improved relations with regulatory authorities.

☐ Facilitation of obtaining insurance coverage for environmental impairment liability.

An EIA is not a one-off process ending with the preparation of a report. It should provide an essential input to project management through a continuing evaluation and re-evaluation of the various environmental issues as project plans are developed and defined. This process should continue throughout the life of the development from conception to final abandonment or closure. It is important, therefore, that careful consideration be given to the scope, management and planning aspects of an EIA. In the management and planning of EIA there are a number of factors that require consideration, as follows (see figure 17).

Terms of reference

In the absence of clear direction an EIA may not cover all the relevant environmental issues but may pursue less important matters. In such circumstances, some of the requirements of the numerous interested parties may not be met. It is therefore important that project management, environmental specialists and representatives of external bodies, who may all hold differing views, meet at an early stage to discuss and agree terms of reference.

This is particularly important where an EIA is required in response to legislative requirements. The terms of reference should define key issues to be covered, decisions to be taken and options to be investigated so as to minimize omissions and the introduction of additional issues in the latter stages of the EIA process.

EIA teams

Multi-disciplinary teams are required to prepare environmental impact statements for a major development or smaller projects of high and diverse environmental sensitivity. Typically, such disciplines should include economics, environmental management, agriculture, forestry, water resource management, atmospheric sciences, etc., or for a small development, an EIA can sometimes be completed by a single person with minimal specialist support. On the other hand, an EIA team may comprise personnel from the developer, the competent authorities and external consultants.

An overall coordinating committee and, in some instances, a management or steering committee should be appointed to:

☐ ensure that all disciplines are working to a coordinated brief, and that information is exchanged and duplication of effort is avoided;

☐ act as the contact point with project management and other interested parties;

☐ identify the need for specialist investigations and their inclusion in the programme;

Figure 17. Principles of managing an EIA

Principle 1: Focus on the main issues
- It is important that an environmental impact assessment (EIA) does not try to cover too many topics in too much detail.
- At an early stage, the scope of the EIA should be limited to only the most likely and most serious of the possible environmental impacts, which could be identified by screening or a preliminary assessment.
- Where mitigation measures are being suggested, it is again important to focus the study only on workable, acceptable solutions to the problems.

Principle 2: Involve the appropriate persons and groups
- Those appointed to manage and undertake the EIA process.
- Those who can contribute facts, ideas or concerns to the study, including scientists, economists, engineers, policy-makers and representatives of interested or affected groups.
- Those who have direct authority to permit, control or alter the project – that is, the decision-makers – including, for example, the developer, aid agency or investors, competent authorities, regulators and politicians.

Principle 3: Link information to decisions about the project
- An EIA should be organized so that it directly supports the many decisions that need to be taken about the proposed project. It should start early enough to provide information to improve basic designs, and should progress through the several stages of project planning and implementation.

Principle 4: Present clear options for the mitigation of impacts and for sound environmental management
- To help decision-makers, the EIA must be designed so as to present clear choices on the planning and implementation of the project, and it should make clear the likely results of each option. For instance, to mitigate adverse impacts, the EIA could propose:
  - pollution control technology or design features;
  - the reduction, treatment or disposal of wastes;
  - compensations or concessions to affected groups.

To enhance environmental compatibility, the EIA could suggest:
  - several alternative sites;
  - changes to the project's design and operation;
  - limitations to its initial size or growth;
  - separate programmes which contribute in a positive way to local resources or to the quality of the environment.

And to ensure that the implementation of an approved project is environmentally sound, the EIA may prescribe:
  - monitoring programmes or periodic impact reviews;
  - contingency plans for regulatory action;
  - the involvement of the local community in later decisions.

Principle 5: Provide information in a form useful to the decision-makers
- The objective of an EIA is to ensure that environmental problems are foreseen and addressed by decision-makers. To achieve this, decision-makers must fully understand the EIA's conclusions which should be presented in terms and formats immediately meaningful.

Source: Based on UNEP, 1988.

☐ programme EIA activities so that the necessary information is available at the appropriate time for permit applications, meetings, etc; and

☐ organize and collate team members' input to produce an EIA document for management, authorities and other interested parties.

Programming

A development comprises a number of phases. For example, preliminary schemes may be followed by feasibility/development studies leading to detailed design, construction and operational phases. A detailed EIA should not commence until the development has reached a fairly well-defined status. However, experience has shown that some environmental input is required during the early stages of a major or highly sensitive project in the form of an initial environmental screening or preliminary assessment. The data obtained during such early investigations would be incorporated in the more detailed EIA. It is important that the programming of the EIA is integrated with other activities so that information is available at appropriate times for the design engineers, relevant authorities, permit applications, etc. This will be the responsibility of the EIA coordinator, requiring close liaison with relevant bodies.

EIA questions

The following questions need to be asked about any major project. All phases of a project life cycle (construction, operation, extension, liquidation) have to be considered:

☐ Can it operate safely, without serious risk of dangerous accidents or long-term health effects?

☐ Can the local environment cope with the additional waste and pollution it will produce?

☐ Will its proposed location conflict with nearby land uses, or preclude later developments in the surrounding area?

☐ How will it affect local fisheries, farms or industry?

☐ Is there sufficient infrastructure, such as roads and sewers, to support it?

☐ How much water, energy and other resources will it consume, and are these in adequate supply?

☐ What human resources will it require or replace, and what social effects may this have on the community?

☐ What damage may it inadvertently cause to national assets such as virgin forests, tourism areas, or historical and cultural sites?

Source: Based on UNEP, 1988.

Sources of information

The availability of information at the appropriate time is important for an EIA if delays and extra costs are to be avoided. Many sources of information are available, such as previous in-house or external experience of developing and operating similar or related installations. Interviews with senior management and the examination of records such as plant designs and operating manuals, previous permissions and consents, plant records (particularly those relating to emissions and effluents), monitoring results and returns from control authorities can all provide useful data.

Public participation

As well as consultations with authorities, early contact should be made with the public, including special interest groups, especially those associated with the preservation of the environment. Public involvement should be an integral part of any EIA system.

Efforts should be made to obtain the views of, and to inform, the public and other interest groups who may be directly or indirectly affected by the project. The authorizing agencies may not always identify the environmental issues which the public perceives to be important and they may also lack the detailed local knowledge that the public possesses.

Advantages of participation may include the provision of information about local, environmental, economic and social systems; the possible identification of alternative actions; an increase in the acceptability of the project as the public will better understand the reasons for the project; and a minimization of conflict and delay. Problems may nevertheless arise. Public participation may, in the short term, be time-consuming and increase costs, and participants may be unrepresentative of the community. In spite of these potential problems, many countries are actively encouraging public involvement in EIA. Not only the public in general, but also workers and their representatives, should participate at appropriate stages in the EIA process.

Cost of EIAs

Initially, EIAs may be expensive to implement, particularly in areas where little is known about existing environmental and social conditions. Design changes produced as a result of EIA findings may also increase capital costs, but it can be argued that the avoidance of deleterious impacts and the maximization of beneficial impacts will outweigh the costs of an EIA system in the long term. The cost of an EIA in-house system will decline once the procedures and techniques have been established and assessment personnel have become accustomed to their tasks. Indeed, thorough investigation of impacts at an early stage of project planning may save money by speeding up the process of implementing a proposal. The costs of EIA are commensurate with the complexity and significance of the problem and the level of detail required. In many countries the cost is borne by the proponent of the development, while in others it is borne by the authorizing body.

In those countries with considerable experience in EIA, the costs may vary at about 0.05 per cent of the project value. This is normally much cheaper than the required retrofitting or environmental liability costs which will have to be assumed if environmental risks have not been taken into account during project planning.

While EIA is an assessment of a project before it is realized (ex-ante assessment), the Environmental Audit (EA) discussed is concerned with already established facilities (ex-post assessment).

3.3. Environmental audit (EA)

EA is a management tool comprising a systematic, documented, periodic and objective evaluation of how well management and equipment are performing in·environmental terms, with the aim of helping to safeguard the environment by:

☐ facilitating management control of environmental practices; and

☐ assessing compliance with company policies, which would include meeting regulatory requirements.

Depending on their needs, companies have developed their own auditing procedures, but the results of international working groups on this topic suggest that there is a consensus on the main characteristics of the steps of environmental auditing (UNEP/IE, 1990b).

The basic steps of an environmental audit (figure 18), developed by the Canadian Naranda Corporation, have been adopted by the International Chamber of Commerce's (ICC) working party on environmental auditing (ICC, 1990).

Figure 18 divides an audit into three parts:

☐ pre-audit activities;

☐ activities at the site; and

☐ post-audit activities.

What is usually audited

☐ Policy, responsibilities and organization.
☐ Planning, monitoring and reporting procedures.
☐ Management and staff awareness and training.
☐ External relations with regulatory authorities and the community.
☐ Compliance with regulations.
☐ Emergency planning and response.
☐ Pollution sources and minimization.
☐ Pollution treatment and discharge.
☐ Resource savings.
☐ Housekeeping.
☐ Land management.

How these activities are carried out is well reflected in the case-studies published on environmental auditing by UNEP's Industry and Environment office (UNEP/IE, 1990b), a summary of which is given below.

Pre-audit activities

In most companies environmental audits are the responsibility of a specialized unit within the environmental department at company headquarters. This independent, centralized unit is also generally in charge of health and safety audits. Smaller components often rely on external auditors.

Environmental auditing needs the strong endorsement and active support of top management. The procedure should be clearly communicated, together with the appropriate incentives. The auditing team has to gain the confidence of the audited units and make clear that the objective is to identify ways to progress, and not to punish managers.

How to select facilities? Most companies who apply this tool audit all of their facilities regularly. One company, for example, audits each refinery every 18 months, medium-risk facilities every three years, and low-risk facilities every five to six years. In each category, facilities to be audited each year are generally selected randomly.

Who should form the audit team? The audit team tends to comprise two to eight people. They may be full-time auditors, subject specialists, representatives from the business unit being audited, representatives from other company plants or qualified external consultants. It is also advisable to include workers' representatives, who will require adequate training and information on the auditing process.

Figure 18. Basic steps of an environmental audit

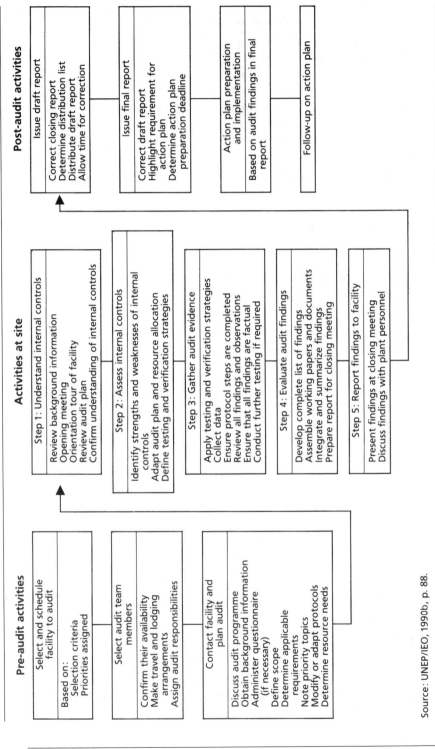

Source: UNEP/IEO, 1990b, p. 88.

Activities at site (conducting the audit)

Each company has developed its own methodology and set of procedures for conducting an audit. Site visits, questionnaires, interviews and review of documents are used. In general, the performance of units is audited according to:

□ government requirements (national, state and local); and

□ internal requirements (corporate, divisional and facility policies, procedures and standards).

In figure 18 the activities at site are divided into five steps: understand internal controls; assess internal controls; gather audit evidence; evaluate audit findings; report findings to facility.

Once the audit has been performed at site, the so-called post-audit activities follow.

Post-audit activities

A draft report should be issued and discussed with the management and workers' representatives of the audited unit. Some companies prefer the audit report to be only a statement of facts, others prefer to include proposals on how to improve things. Once a final report has been issued and adopted, an action plan based on the audit findings has to be prepared, usually by the audited unit assisted by the respective specialists. Proposals for improved environmental performance have to be implemented by the department in charge of the audit.

To improve confidence, at both local authority and community levels, in a company's operations, information should be periodically supplied about audit results and measures taken to improve environmental performance.

3.4. Organization of environmental functions

What organizational structure does an enterprise need to include effectively environmental considerations in all areas and fields of management?

□ Is a full-time environmental manager, whose responsibility to the firm is the environment, a viable solution?

□ Do we need an environmental coordinator to try to bring all those people together who could be involved in environmental questions in the company, and coordinate their actions?

□ Is an environmental committee a feasible solution?

☐ Who represents the enterprise in environment matters to the local authorities and the community?

☐ Which board member should be made responsible for environmental action?

☐ What are the relative advantages of environmental specialization versus integration in existing units?

☐ Are there any legal requirements for an "environmental chargé d'affaires"?

The way the environmental protection function is organized in a firm depends principally upon the size of the firm (see UNEP/IEO, 1991). Small, medium and large enterprises all have different formats. The simplest, of course, is that of the small-scale enterprise, which can be defined as an enterprise in which one person makes all the management decisions. In such firms the environment manager is the person, usually the owner, who performs all the other management tasks.

Medium-sized firms with, for example, 50 to 300 employees, have their own problems when it comes to setting up an environmental protection function. They cannot add the management load to the managing director's existing job, as can small-scale firms. In some cases they may be able to employ someone to work exclusively as the environmental manager/engineer, but in most cases this task is simply added to someone's existing responsibilities. It is here that the difficult questions start. Who should this person be? What level of seniority? What sort of professional background? What training will he/she need? What will be the relationship with the operations people?

Many companies believe that whoever is assigned this task should have a technical background in either engineering or, preferably, chemistry and should have had actual experience in operating the plant. The reason for this is that the environmental manager will be dealing primarily with the operations managers and will need such a background in order to do so effectively.

Of 32 senior environmental executives in leading European companies, James and Stewart (1994) found that most were male, had scientific or technical qualifications, little formal business or environmental education, had transferred from research or production positions and had a high level of job satisfaction. However, many were frustrated by resource constraint and a sometimes slow pace of change in their companies.

Often an environment protection function is added to the occupational safety and health unit. There are good reasons to do so (see section 2.9). This is confirmed by James and Stewart (1994), who found that in half of the companies in their sample the most senior environmental executive combined environmental with operational OSH responsibilities. Industrial engineering units are already in charge of process optimization and productivity in an enterprise and, thus, are also well qualified to take up environment protection functions within the enterprise.

Organizational forms

A feasible organization of environmental responsibilities and functions in a medium-sized company may be as follows: the technical director has the overall responsibility for environmental protection. This person chairs an environmental committee where, as a minimum, plant chiefs, maintenance, and OSH units are represented. The maintenance or OSH units are responsible for carrying out the operational tasks of environmental protection. This is a "minimum" approach which does not guarantee that environmental considerations are included in all fields of management, as proposed in Chapter 2. To ensure that, there may also be an environmental committee at the department level representing marketing, R&D, products, personnel, financing and legal advisers.

In large-scale corporations James and Stewart (1994) have identified three organizational forms of environmental management:

- **The central coordination model**, which involves a relatively small head-office unit with responsibilities for policy development and coordination, specialist advice and relationships with legislative and other stakeholders. This head-office unit works with environmental units in autonomous divisions which are responsible for the day-to-day tasks of environmental protection.
- **The central departmental model**, in which a central department has substantial dealings with sites and has considerable operational responsibilities. In such a central department, the company would house all the needed specialists such as chemists, physicists, biologists, hydrologists and so forth. This central department would be responsible for operational tasks, for example monitoring of emissions, and would provide in-house consultancy services and manage projects across business units.
- **In a business unit based model**, business units have the primary environmental responsibilities, but their executives handle some matters on behalf of their parent company; for example, group responsibilities may be divided between business unit coordinators.

In all organization models, however, line managers have an important role to play. Line management's responsibilities include the implementation of projects and programmes, controlling production operations, products and services, and ensuring that all activities within their domain are compatible with stated environmental policies.

Effective organization of environmental protection

☐ A member of management is formally designated as the leader or coordinator of environmental protection programmes.

☐ Sufficient human resources are available for implementation of environmental protection programmes.

☐ Management structures are such that an adequate flow of information on environmental matters reaches senior management.

☐ There is an internal review programme designed to identify and evaluate environmental hazards at regular intervals.

☐ There is an effective routine flow of environmental information among all appropriate levels within the facility, including:

– through formal mechanisms, i.e. line management and committee structures;

– through inclusion in site publications, technical meetings and other general mechanisms for information flow.

☐ Management, supervisors and staff are aware of the environmental issues which are relevant to their areas of responsibility.

☐ All staff have an appropriate level of awareness of regulatory requirements.

☐ Employees responsible for managing and implementing environmental protection programmes are trained and qualified to the appropriate level.

☐ Where necessary, site guidelines detailing particular environmental sensitivities, regulations, etc., are available to all staff.

☐ Environmental issues are included in regular reporting to business headquarters and an annual report on environmental matters is prepared.

☐ Records and reports of incidents, monitoring data, waste disposal documentation and other environmental information are maintained in a well-organized and accessible form, and are available to the environmental coordinator.

☐ There are procedures to ensure that:

– written operating instructions for major operations contain the necessary up-to-date environmental protection instructions;

– regulatory constraints are incorporated in operating instructions.

☐ Programmes are implemented which give an appropriate background of environmental awareness for all employees.

Source: Based on UNEP/IEO, 1990b, p. 41.

Among the tasks of environmental units at corporate and business-unit level is the establishment of contacts within the enterprise and with central government, specialized agencies, politicians, pressure groups, employers' organizations and trade unions, specialized research institutes, the media, and others who have an opinion about and concern for environmental issues.

These concerns must be translated into corporate actions, and the external groups informed of the actions taken. Thus the units act as co-ordinating groups which will have close links with the board of directors, since many of the issues considered may have an important impact on the future direction and strategy of the company. Many large corporations have introduced a board-level environmental committee to advise the board of directors. In times of emergency the full support of top management is required to avoid an escalation of an environmental issue or problem into a major crisis.

The effectiveness of an environmental coordinating group or an executive officer will be reduced if environmental issues are not included in the decision-making process.

The environmental executive may also act as a catalyst by increasing awareness and arranging training courses for line managers in environmental subjects, including permit applications, external contracts and emergency response. A primary responsibility of top management is to encourage the environmental skills of managers and progressively guide the values and cultural attitudes of the enterprise towards greater environmental sensitivity and concern over the external dimensions of the job.

The bodies an environment manager or coordinator has to liaise with are shown in figure 19.

Environmental management as a career step

To demonstrate the fact that an enterprise is taking environmental issues seriously, the post of environmental manager could be seen as a career step towards senior management positions. Below are several good arguments for doing so.

☐ To get results in this job, particularly if the organizational linkages to the other departments are (deliberately) somewhat ambivalent, the appointee will have to demonstrate the principal characteristics of senior management:

– communication skills, ability to communicate a vision, persuasiveness and leadership. Without these skills the person would not be promoted to a senior management position.

☐ If the person has the potential for developing these skills, this post provides an excellent on-the-job training ground for doing so.

☐ The fact that a future senior manager is spending some time in "environment" will demonstrate to the whole enterprise that top management considers this function to be important. This will help attract higher-quality people, and the fact that top management will learn more about this function and how it operates can only be beneficial to the firm.

Figure 19. Contacts of an environmental coordinator

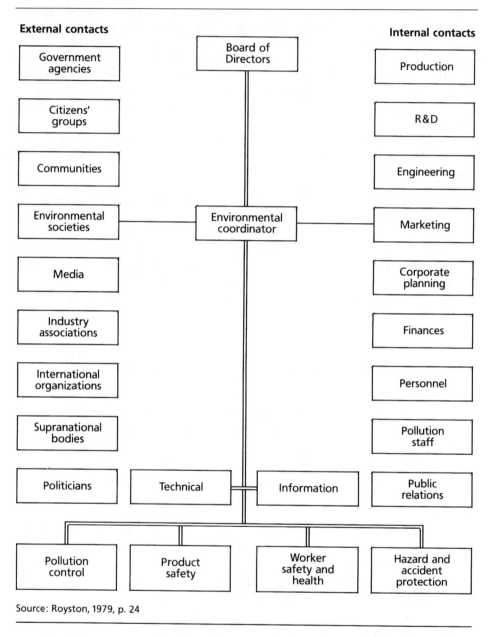

Source: Royston, 1979, p. 24

Following these general considerations on how to organize environmental functions in an enterprise, let us now look into some concrete cases to find out how some enterprises have tackled the problem.

Company cases

The environmental effort of a Canadian aerospace, electronics and engineering group is headed by a Vice-President who reports to the corporate Senior Vice-President of technology (see figure 20). The corporate environmental affairs department is divided into six units headed by directors, who meet approximately once a week to discuss policy and mutual concerns. The department is responsible for policies, the overall programme and coordination, environmental planning, the major assurance programme and review of environmental financial matters, training, inspection and follow-up.

The operating company, sector and plant environmental staff (composed of basically the same six functions) have responsibility for the day-to-day hands-on operation (UNEP/IE, 1990b).

A German multinational chemical enterprise has integrated environmental protection into its organization in the following manner (see figure 21).

One board member is in charge of environmental protection and occupational safety and health. Together with two other fellow board members who are in charge of R&D and personnel, the board committee on environmental protection and OSH is formed. This committee is the supreme decision-making body for related questions worldwide. Another committee, which consists of board members and the heads of the central works administration and engineering, develops the overall technical concepts. The company is further organized into product-oriented business units and function-oriented central units. There is one central unit dealing with works' administration, environmental protection and OSH which has respective counterpart departments in the business units. The central R&D unit is also concerned with environmental matters in its sections dealing with safety of processes and facilities, energy and process control technology.

Apart from defining environmental management in the organization, it is important to ensure staff commitment and participation. That is why some companies encourage their workers to make proposals concerning the environment through the existing suggestion systems. Other companies have opted to create their own "environmental suggestion scheme". Success is also due to initiatives which include environmental concerns in the work of quality circles. These aspects will be dealt with in the following section on communication and participation.

3.5. Communication and participation

The following is an analysis of a classic pattern of communication between a company and the community. Local Green activists claim to have discovered that a fertilizer producer has been releasing toxic substances into the soil for several years and therefore contaminating the groundwater. The

---

Effective external relations

☐ There is an effective and clearly defined system for liaising with regulatory authorities at the local, regional and national levels.

☐ There is an effective and clearly defined system for liaising with the press and other media, local community and interest groups and the general public, using the services of public affairs specialists where necessary.

☐ Local communities and interest groups are adequately informed of the environmental impact of the installation, using environmental monitoring data when appropriate.

☐ Complaints are investigated and recorded in a systematic fashion and appropriate actions taken.

☐ Where necessary, there are systems for alerting the public of imminent hazards.

☐ There is an appropriate level of participation in projects promoted by local or national bodies.

Source: Based on UNEP/IEO, 1990b.

---

spokesperson of the enterprise denies all charges and states firmly that everything has been done to ensure a safe production process. The Green activists take soil and groundwater samples, have them analysed by a laboratory and present the results in a press conference. Indeed, contamination of the groundwater has been found, the concentration exceeding threshold values. The company spokesperson alleges that owing to an error in cleaning, contaminated water may have leaked into the soil. The public is assured that this was the only accident that has happened and that everything is under control. The company presents an independent expert, who criticizes the measuring methods of the laboratory and states that the contamination, though threshold values had been surpassed, is not at all harmful. Green activists press local authorities to visit the factory grounds and investigate the matter. From this it becomes obvious that toxic substances had been leaking into the groundwater for quite some time during the manufacturing process. It is only then that the enterprise admits that some leaks have occurred. Despite this admission, the enterprise states that the toxicity found in the groundwater was not scientifically proven.

The enterprise states it they will take "voluntary" action to ensure a complete encapsulation of the criticized process. The local Green activists are not satisfied with this result and try to find other areas in which to attack the company.

So far a classic communication pattern. As a worker would you be proud to work in this company? As a citizen how would you react if the company planned to extend its prodution facilities? As a consumer would you buy the products of this company if there were alternatives on the market?

Figure 20. A Canadian aerospace/electronics/engineering group

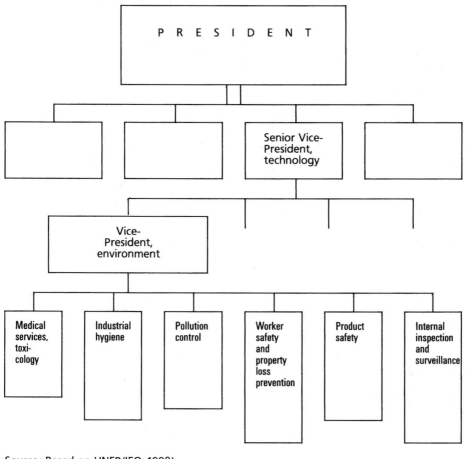

Source: Based on UNEP/IEO, 1990b.

The chemical industry in particular has learned how difficult it is to restore credibility after a number of incidents, and even though, for example, the chemical industry in Western Europe has dramatically decreased its effluents, the polluter image persists.

It is understandable that people living or working near a chemical plant are worried about effluents, fumes, and worse – worries which the events in Seveso, Bhopal and Flixboro have done nothing to alleviate. Consequently, the public relations efforts of firms in these industries must be dedicated largely to explaining clearly and openly the precautions which have been

Figure 21. A German multinational chemical enterprise

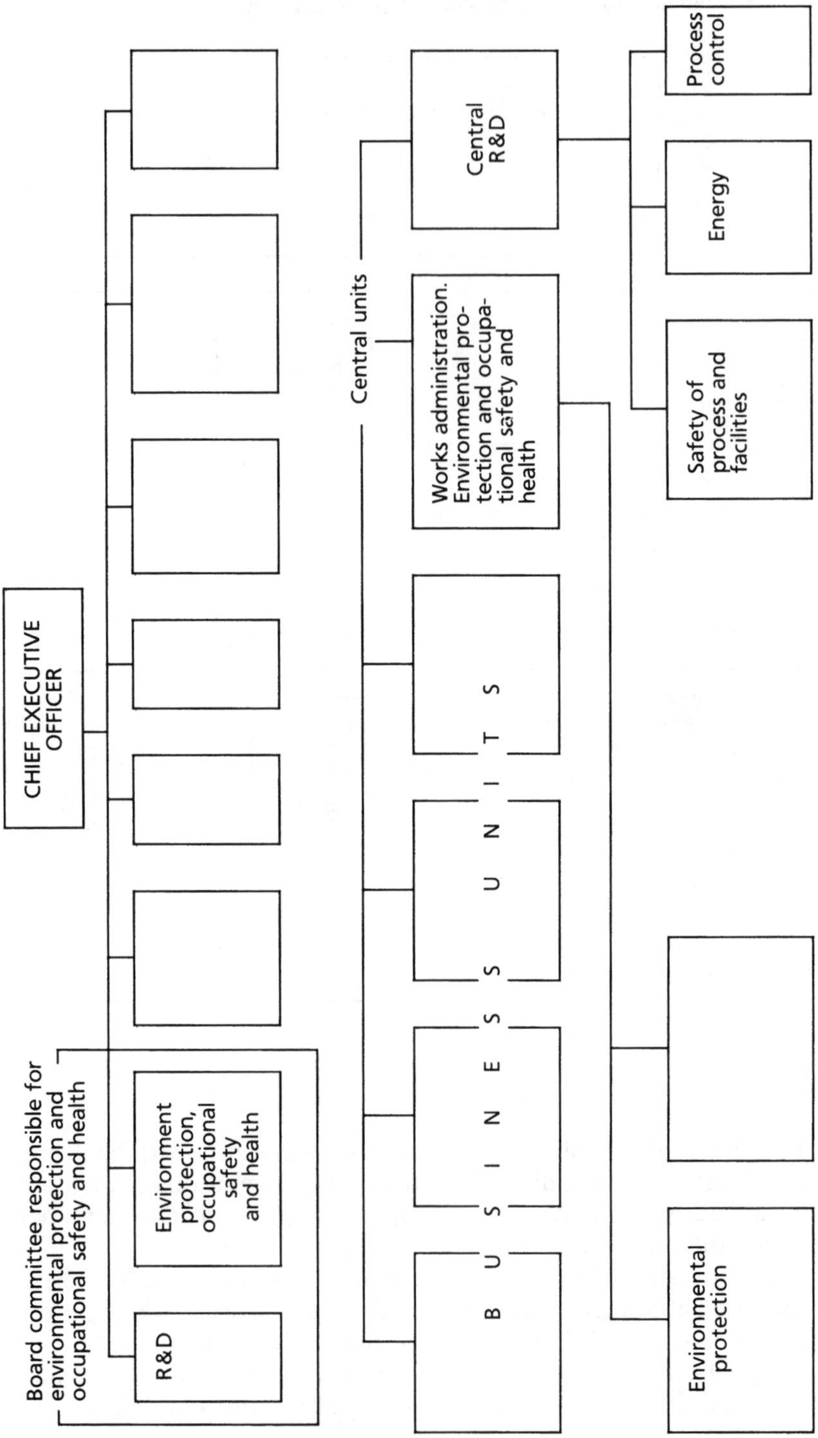

taken to prevent any of these unpleasant occurrences. In the past, the tendency was for firms to remain silent on these matters or to react as described above, and hope that nobody would discover what products were being processed, what risks were involved, and what precautions were being taken. This is now widely recognized as a mistaken approach.

Overcoming a polluter image

What can an enterprise do to overcome a polluter image and the resulting public resistance?

The first course of action is to take effective measures to protect the environment. The second is not to try to appear Greener than the enterprise really is, since it will undermine a company's credibility as we have shown in section 2.2.

These two considerations are fundamental for all further public relations. Concerning ongoing operations, many companies:

☐ issue information brochures;

☐ organize round-tables;

☐ issue invitations to factory visits;

☐ participate in information activities of the local Employers' Association or Chamber of Commerce;

☐ have regular discussions with pressure groups;

☐ liaise with local authorities and the press;

☐ financially support environmental groups and encourage staff members to participate;

☐ establish information programmes for different target groups of the community such as schoolchildren, students, the working population, specific professional groups (e.g. medical doctors), women, retired persons.

If the policy of broadcasting the "good" actions of an enterprise succeeds in convincing the community, it will bear fruit when that company wants to implement a new project which requires approval by national, regional or local authorities and which takes into account public opinion. Figure 22 shows possible events in a project, and solutions to meet the interests of the parties involved.

Figure 22. Communication techniques (assisting in the search for a consensus)

Communication characteristics — Corporate planning objectives

1 = Low   ▓ = Capability
2 = Medium
3 = High

Level of public contact achieved	Ability to handle specific interest	Degree of two-way communication	Public communication techniques	Inform and educate	Identify problems and values	Get ideas and solve problems	Feedback	Evaluate	Resolve conflict and reach consensus
2	1	1	Public hearing		▓		▓		
2	1	2	Public meetings	▓			▓		
1	2	3	Informal small group meetings	▓	▓		▓		
2	1	2	General public information meetings	▓					
1	2	2	Presentations to community groups	▓					
1	3	3	Information coordination seminars		▓	▓			
1	2	1	Operating field offices		▓	▓			
1	3	3	Local planning visits		▓	▓			
2	2	1	Community survey research			▓			
2	2	1	Information brochures and pamphlets	▓					
1	3	3	Field trips and site visits	▓					
3	1	2	Public displays	▓					
2	1	2	Model demonstration projects	▓			▓		▓
3	1	1	Material for mass media	▓					
1	3	2	Response to public inquiries	▓					
3	1	1	Press releases inviting comments				▓		
1	3	1	Letter requests for comments			▓			
1	3	3	Workshops		▓	▓			▓
1	3	3	Advisory committees		▓	▓			
1	3	3	Task forces		▓	▓			▓
1	3	3	Employment of community residents		▓				▓
1	3	3	Community interest advocates		▓	▓		▓	
1	3	3	Ombudsman or representative		▓	▓			
2	3	1	Public review of impact statement	▓	▓	▓	▓	▓	

Source: Boland, 1986, p. G4.27.

Action guidelines to minimize conflicts

☐ Analyse the scene.
☐ Understand the underlying issues.
☐ Establish your objectives.
☐ Contact a broad cross-section, including workers.
☐ Identify the opinion-makers or decision-makers.
☐ Set up a joint planning mechanism.
☐ Hold meetings on neutral ground.
☐ Listen rather than speak.
☐ Don't accept offices, e.g. chairman or secretary.
☐ Allow an acclimatization period.
☐ Be open, frank and sincere.
☐ Keep a low, but level profile.
☐ Offer guarantees; accept commitment.
☐ Remember that "process" is as important as "substance".
☐ Keep the media involved; issue news-sheets.
☐ Help other decision-makers maintain their constituents' trust.
☐ Don't underestimate the other participants' technical knowledge.
☐ Be positive.
☐ Explore every way of doing what the others want.

Source: Boland, 1986, p. PM2.13.

There are a number of public communication techniques which assist in reaching a consensus. In figure 22 these techniques are evaluated according to the level of public contact achieved, their suitability in handling specific interests and the degree of two-way communication.

Environmental reporting

Corporate environmental reporting is increasingly becoming an important tool for public communication. According to a KPMG environmental reporting survey of the *Financial Times* top 100 companies in the United Kingdom, 20 produced a separate environmental report in 1993, and 34 in 1994; these numbers are increasing rapidly. The pioneering firms were those which wanted to demonstrate their social responsibilities. They were followed by companies in environmentally sensitive industrial sectors which had often experienced heavy public criticism and are hence trying to document and communicate their environmental efforts in order to improve their public image (Baumert and Hermann, 1996). But many companies still have to learn how to go about environmental reporting. Their reports should be neither a collection of case-studies nor a public relations document illustrating a company's commitment to the environment. Environmental reports should be targeted to the information needs of the different stakeholders and should credibly report environmental activities relating the present situation to benchmarks to be achieved.

Ten Brink et al. (1996) propose four stages of a new environmental reporting approach developed for IBM. The first step is to identify the important external and internal stakeholders, and determine their key concerns and expectations regarding the corporation's environmental performance. The second step is to establish which environmental performance parameters should be addressed in the environmental report and the stakeholder's priority ranking of these parameters. Thirdly, the performance of the corporation has to be assessed against these parameters, and fourthly, an environmental performance profile should be prepared and the results communicated back to the stakeholders.

The European Environmental Management Audit Scheme (EMAS) also stipulates environmental reporting: "An environmental statement shall be prepared following an initial environmental review and the completion of each subsequent audit or audit cycle for every site participating in this scheme". According to EMAS, the environmental statement shall include: a description of the company's activities at the site in question; an assessment of all the significant environmental issues of relevance; a summary of the figures on polluting emissions, waste generation, consumption of raw materials, energy and water, noise and other factors regarding environmental performance; a presentation of the company's environmental policy, programme and management system implemented at the site in question; the deadline set for submission of the next statement and the name of the accredited environmental verifier. (Regarding requirements and cases of environmental reporting, see also UNEP and Sustain Ability, 1994).

Staff commitment and participation

But how can the general public be convinced of the "environment friendliness" of a company's operations if its staff is neither convinced nor motivated to contribute to environmental protection?

Looking into the ways of committing the staff to the environmental cause, there should first be a clear policy statement followed by corporate environmental action which is backed by top management. Secondly, changes need to be well prepared. Here Winter (1988, pp. 78 and 79) has established some useful guidelines:

☐ Prepare decisions on in-house environmental protection measures well in advance by providing information in good time and motivating members of the management team.

☐ Before taking a decision on measures which could be controversial, wait for an opportune moment, e.g. the announcement of a successful year's trading, a special anniversary or a major order.

☐ Continue to work on members of the management team who have been outvoted on an environmental protection issue, rather than treating them as defeated opponents.

☐ Give wide in-house publicity to positive decisions on environmental protection measures, to ensure that the management team is seen to be acting in unison and to bind its individual members more strongly to the collective decision.

☐ Where there is a positive response to a collective decision on environmental protection measures, stress the involvement of all concerned (and thus let everyone share in the success), both in the company and in the media.

☐ Where there is a negative response to a collective decision on environmental protection measures (e.g. complaints about the quality of recycled paper), stand by whichever members of the management team are under fire and thus strengthen team spirit.

☐ Take every opportunity to thank and praise members of the management team for their environmentalist attitude, with a view to motivating them for future measures.

In simultaneously committing and unifying management, staff members have to be convinced, motivated and provided with adequate means of participation. Using as a basis his own practical experience as an entrepreneur, Winter proposes first arousing curiosity and interest among staff, and then continuing by gradually raising the level of knowledge (environmental competitiveness) within the company, using deliberate propaganda action to develop environmental awareness, followed up by encouragement and practical guidance concerning desirable behaviour at the workplace, at home and during leisure activities.

Many enterprises have integrated environmental considerations into their suggestion schemes or quality circles. Other companies have opted to create additional small group activities focused on environmental protection. 3M's PPP (Pollution Prevention Pays) programme owes its success, to a great extent, to staff involvement.

Participation also means involving the elected workers' representatives. The works council may designate a member to represent workers' interests in environmental matters and to demonstrate that not all initiatives are left to management.

The 3P (Pollution Prevention Pays)
suggestion scheme of the 3M company

Any technical employee, or group of employees, can nominate a project for 3P by filling out the standard submittal form. The application is then reviewed by the 3P Coordinating Committee made up of representatives from all the technical groups (engineering, manufacturing, and research and development). Thus, judging is by a group of one's peers. The Coordinating Committee meets quarterly to discuss the entries from all 3M's American locations.

Once a project is approved, managers at the operating division for the project identify the personnel who made direct, measurable contributions. Managers and supervisors are recognized only if they were involved in the project hands-on. Environmental Engineering and Pollution Control personnel are generally not eligible. Awards are presented by division managers at a special luncheon or other important division function.

To keep the programme up front as a priority for the technical people, the company runs 3P information exchange seminars company-wide to share information about successes in various groups. Also, 3P laboratory workshops are held to emphasize pollution prevention in new product design, where the employees' written material (including external public relations materials) that tells the story of 3P is shared.

For both internal and external use in technology transfer, regular idea sheets are printed, one-page summaries that describe the methods and results of 3P projects and identify the people involved. A 3P brochure, a 3P sound/slide show and a pollution prevention video have also been provided.

Source: Based on Bringer and Benforado, 1989, p. 20.

Workers should gain personal advantages from environmentally friendly behaviour. There may be a special bonus within suggestion schemes including environmental protection. Incentive schemes may remunerate the saving of energy (if this can be influenced by the worker), reduction of waste, or any other measurable improvement of environmental performance.

Using the great variety of communication and participation techniques helps to ensure that those who live or work near their factories look upon them as providing benefits to the region. Having a reputation like this can provide many benefits for the firm such as making it easier to recruit workers, speeding up the granting of permission to construct new buildings, and so forth.

3.6. Training for environmental management

"Training forms part of the success of our enterprise. Our responsibility is to make environmental protection a main topic." This is the key text of a half-page advertisement of a European multinational chemical company published in a leading Swiss newspaper. This proves that environmental training becomes an advertising object to convince the public that competent and responsible staff are caring for the environment.

Environmental management training has introduced environmental issues into the mainstream of decision-making by imparting knowledge, creating attitudes, and teaching skills which will enable managers to recognize and fully consider the environmental consequences of their actions (Royston, 1989). Such training must be relevant to the current problems of both the enterprise and the community. It can help in seeking creative alternatives, and in anticipating and dealing with potentially difficult environmental situations. While many environmental issues and problems are common to both developing and industrial countries, their solutions and procedures may vary. Environmental education, therefore, should be based within the economic and social context with which the managers are familiar. Environmental management training encourages policy formulation and decision-making which avoid confrontation with the community on environmental issues.

It can help managers to communicate with technical staff about the need to deal with environmental problems in their search for cost savings or the generation of new income for the enterprise.

Targeting environmental training

In all training programmes the target group must be identified and the programme designed accordingly. The basic question to be asked is: What skills are required for environmental management? According to a British study (ECOTEC, 1990), the actual tasks associated with environmental management include:

☐ maintaining an awareness of environmental legislation;

☐ developing process changes for compliance with environmental legislation;

☐ monitoring and analysis of waste streams; and

☐ ensuring conformity with environmental standards.

Senior managers, including plant chemists, chief scientists and technical/plant directors, will typically require a science or engineering degree combined with at least five years' industrial experience. At these occupational levels, the following main sets of skills are required:

☐ to act proactively to identify environmental opportunities and threats;

☐ to appreciate and understand environmental legislation;

☐ to communicate the environmental challenge to staff, clients and the community;

☐ to assess the environmental impact of the plant and to ensure conformity with environmental standards; and

☐ to determine optimal technical solutions to the need for environmental management.

Examination of the manpower implications of new environmental legislation and technical change suggests that the requirement for these skills will increase.

In their study on environmental training in British and German companies, North and Daig (1996) found that the desire to attain accreditation according to EMAS or BS7750 was an important driving force for environmental training at all levels in a corporation.

In the training of *production managers,* topics such as the environmental impact of projects, plants and processes, layout of material storage and the transport of materials and products in and out of the plant should be dealt with. All are part of the production manager's job, and each constitutes a potentially significant hazard for the local community.

Supervisors and operatives require few pre-entry qualifications. The basic skill requirements are production/process engineering skills. The increasing devolution of responsibilities to more junior staff, associated with the introduction of total quality management, also suggests that more environment-related skills may be needed by supervisors and operatives.

Environmental management tasks for operatives generally relate to the operation of plant and equipment and, increasingly, to sampling and monitoring of waste streams. More senior operatives may be involved in the collation of environmental data and may take responsibility for on-the-job training.

There are several possibilities for acquiring the necessary skills. In the above-mentioned British study managers responded that they increase their skills and awareness through short courses and information in trade journals, and by attending exhibitions. Seminars and conferences are seen as a valuable source of information and distance learning is acknowledged as a flexible training medium. The current main deficiency in training provision is a lack of properly targeted and specialized short courses. Some training is available from the suppliers of new equipment as part of their sales package.

Developing training programmes

For firms who want to set up their own environmental training or for training institutions desiring to capture part of the new environmental training market, there is always the question whether to integrate environmental topics into existing courses or to develop specialized environmental courses. Certainly, both alternatives can be viable.

Table 11. Integration of environmental material into training courses

Existing content	Extra environmental material
Process of management Human relations and communication The production function Work study: Its purpose and practice Production control: Its scope and importance Quality control Maintenance: Achieving effective maintenance Basic cost Inventory control: Its purpose Labour cost controls Purchasing management Personnel management and development Sales planning Incentives Industrial relations	Systems approach to sustainable development Communication, participation in environmental matters Raw materials, residues and clean technologies make best use of production facilities Energy accounting, non-waste technology Environmental aspects of waste reduction Cost of inefficiency, leaks of energy and material Externalities Residues control and scope Utilization of residues and renewable resources Social responsibility, worker health and safety Meeting basic needs with new products Incentive schemes based on energy saving, material saving Better environment for workers and their families, union involvement in environmental protection Environmental management Environmental auditing

Source: Adapted from Royston, 1989, p. 58.

The benefits of *integrating environmental education* with other management subjects are considerable (Royston, 1989). This integration reduces the source of conflict in the mind of trained personnel and better equips them for decision-making in the real world (see North and Daig, 1996).

The most successful format for this kind of environmental management training is short self-contained modules which deal with the environmental aspects of major topics taught during the course. This is seen as more effective than examining the environmental dimension of management as a component of a longer course.

The ultimate aim is to heighten the environmental awareness of all those involved in management teaching, so that a consistent environmental

Figure 23. Content of university postgraduate courses in environmental management, 1990

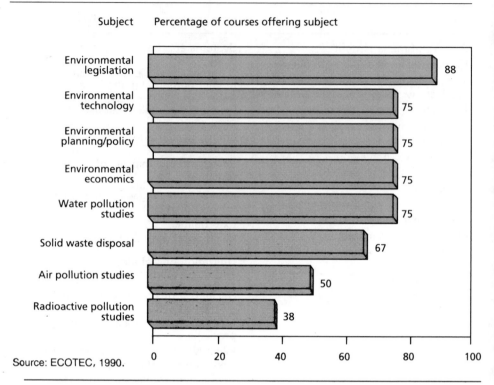

Subject   Percentage of courses offering subject

Source: ECOTEC, 1990.

thread runs through all curricula. Whilst this may be unrealistic at present, it serves a purpose as an overall goal at which to aim during the evaluation of management training programmes and in the development of new courses and curricula.

Integrating environmental and other dimensions of decision-making into formal, lecture-based training programmes and game-playing exercises is difficult. However, the introduction of the basic concepts of environmental management alongside other management functions may encourage or assist trainees themselves to better integrate the different subjects being taught. The presentation of case-studies in which environmental issues have a central role is another way of demonstrating the indivisibility of managerial decision-making.

Table 11 shows, for example, how environmental topics could be included in an existing production management programme. The left column lists the content of the existing training programme, while the right column indicates the environmental material which could be integrated into a new management programme.

Despite the advantages of integrated environmental education, a case exists for *training courses on specific topics.* In a short course, full use should be made of locally relevant material related to the industrial sector and geographical location of the trainees together with presentations on local needs and resources. The basic need is to put forward the positive aspects of environmental management and the contribution they can make to good management practice. Thus, to a group of marketing managers this means ways of marketing safe products; to production managers it means avoiding waste, cutting costs and finding and using new resources; while to personnel managers it means the need to promote the optimal and rational use of resources, including human resources.

In one- or two-day courses it is not possible to cover the full range of topics. However, the use of audiovisual techniques and films provides an effective way of communicating multidimensional aspects of the environment.

Simulating an environmental impact assessment (EIA)

The presentation of a simulation exercise might start with the description of an actual industrial, agricultural or urban development proposal. Documentation about the proposed development would then be provided and the trainees given the opportunity to ask questions. Groups of trainees would analyse the different stages and aspects of the proposed development with a view to preparing a list of project activities which might have a significant effect on the environment. Each group would then be given a dossier listing the physical and social parameters of the environment in which the project would be located. At this stage, site visits could be suggested to provide additional data about the environmental components that might be affected. The opportunity might also be taken to meet local residents to obtain their views about the parts of the environment important to them and their significance to the local community.

Using the information obtained on project activities and environmental components and processes, trainees would be invited to prepare a preliminary EIA with a view to identifying those aspects of the environment which might be significantly affected. This might include a preliminary assessment of those environmental areas in need of protection and the necessary modifications to the project to minimize environmental damage. The preliminary assessment prepared by each group could then be considered at a simulated public hearing, where trainees play the role of various community members, such as farmers, fishermen, shopkeepers, businessmen and local officials. What emerges at this stage is a reassessment of environmental priorities and the establishment of links between the environment and development objectives. After the role-playing of the public hearing, the project proponent and local community members may be invited to the training centre to hear the summary of the case and to comment on the analysis and findings. The learning experience of such a simulation can also be obtained by using fictitious development proposals, but without the authenticity of an actual proposal.

Source: Based on Royston, 1989.

There are now also a number of postgraduate courses that exist in environmental management. Course subjects offered according to the above-cited British study are displayed in figure 23.

Training methodology

It is not just that environmental studies need to be acquired; the more demanding task underlying environmental management training is to change attitudes. An effective and adaptable training programme will be based on active learning, which involves doing rather than passive listening.

A proven mechanism for environmental management education is the simulation of an EIA integrated with role-playing, field visits and lectures which elucidate technical and managerial learning points. Beyond this, it is possible to develop environmental games. In general, they facilitate learning through role-playing, but without field visits, involvement of local community members or project proponents, situations remain artificial. However, simulation games offer an effective, low-cost and efficient approach to achieving learning objectives. This, combined with the opportunity for trainees to undertake field studies, increases the interest in, and the value of, simulation exercises.

A training method which can be effective in attitude change is the one which the ILO uses under the name of Flexible Learning Packages (FLPs). Although they are not intended for conveying in-depth knowledge, they can be useful as an introduction to the subject. The ILO has prepared a number of FLPs on the subject under the title *Environmental management training* (Boland, 1986; see also Evan-Stein, 1990).

Training and trainers

To achieve integrated environmental management training, priority must be given to training trainers so as to increase their knowledge, management skills, and sensitivity to environmental education, but training should also be extended to cover broader interest. On the one hand, the different methodologies required by environmental issues should be covered, particularly those needed to bring about a change in attitudes; on the other hand, the requirements of specific subjects and the need to obtain access to relevant background material on the environmental dimension of different areas of management expertise should be made explicit.

Finally, each of the training sessions should include an action plan. This should be considered under three headings.

☐ First, each trainer should explicitly state why there is a need to introduce the environmental dimension into the training session, and what will be introduced. It is also useful to consider what material or methodology currently used is being retained or discarded because it does or does not promote environmental protection and good environmental management. Similar attention should be given to the introduction of new material.

☐ The second stage is concerned with how trainers intend to teach and includes the adoption of different approaches to convey an effective environmental message.

☐ The third stage is concerned with when the action plan and various environmental issues will be introduced into the ongoing training sessions, and when a timetable transforming the existing approach will be implemented.

In many training establishments it has been found that the why, the what, the how and the when of introducing the environmental management dimension into the training programmes, linked with what should be initially taught, provide a useful framework for trainers and their training courses.

3.7. Dealing with wastes and pollution

The production of industrial goods involves the extraction of natural resources, their utilization in the manufacture of industrial products, and the disposal of unwanted materials not utilized in the final product. These processes raise two major environmental problems for industry (UNEP, 1989b).

The first is how to dispose of the wastes that industry generates: gaseous waste causes air pollution, and liquid wastes pollute ground and surface water supplies. In the past, many wastes were never treated at all. Today, inefficient treatment processes pose many problems. Incineration, at sea or on land, and the dumping of wastes into the oceans, are under severe attack.

The second major problem faced by industry is how to conduct its affairs without subjecting its workers and the public to risk – risk of accident, risk of exposure to dangerous chemicals, and risk in the workplace.

It is not only production that causes waste and pollution, but also the disposal of products, which is an equally important problem. In some countries industry is facing regulations to recycle products and packaging material. We are approaching a time when car dealers will soon be obliged to take back cars for recycling or supermarkets obliged to collect packaging materials.

The size of the problem is enormous. In the United States alone in 1990, approximately 180 million tons of solid waste was generated by homes and commercial establishments of which only about 13 per cent was recycled (Nulty, 1990). Throughout the world there are breweries, textile plants, sugar plants, tanneries, meat and fish processors, pulp and paper plants, dairies and chemical plants discharging more or less treated wastewater sometimes with high biological oxygen demand, all polluting rivers and lakes. Gaseous emissions are widespread. In the OECD countries hazardous wastes are produced at a rate of 300 to 800 million tons a year, or even more.

Waste management

According to Balkau (1990), modern environmental management methods have now given us the ability more than ever before to avoid many of the problems which were earlier accepted as inevitable, but which are now seen to impede national development. Such models emphasize:

☐ low-waste production processes;

☐ the use of safer processing chemicals;

☐ maximum recovery of residues;

☐ cost-effective treatment of residues;

☐ preventive maintenance;

☐ good housekeeping; and

☐ competent operation.

Prior assessment of waste generation and of environmental risk allows systematic identification of priority problems so that the most cost-effective action can be brought to bear on the problem.

Such an approach is quite different from the earlier "end-of-pipe treatment" mentality which let production processes generate as much waste as they wanted, and then concentrated on building a treatment plant for whatever was produced. This approach was expensive, and often showed little real benefit to the environment. In many cases it did not actually reduce pollution, but merely transferred it to another environmental sector, as for example by concentrating pollutants in treatment sludge which was then disposed of onto the land.

The new approach attacks the environmental problem at its roots. Rather than focus on a single technique in the hope that it will solve all problems, it applies a systematic set of techniques adapted to each situation in question.

The new approach also acknowledges that waste management is a cooperative activity to which both the government and the industry sectors must contribute equally. Without a clear government policy and sensible regulations (effectively enforced), there is no predictable framework which industry can use for planning. Conversely, without the use of modern environmental management tools industry cannot take a proactive approach to environment, and is often unable to comply with the ever-tightening regulations.

As the "polluter pays" principle is now widely applied and the amount of waste is increasing, waste disposal is becoming more and more expensive for enterprises. On the other hand, waste itself has become a huge business.

Individual companies make profits by selling their wastes or reduce production costs, e.g. by heat exchange, by reuse of treated waste water or by reducing gaseous emissions.

So what are wastes and how can we deal with them in an environmentally responsible manner?

Wastes can be virtually anything and can be of any consistency, whether solid, liquid or sludge. It is sometimes argued that we should distinguish between wastes and recyclable materials, but in practice this is difficult as the state of the art and the "waste" market changes rapidly. Hazardous wastes, according to a definition by the World Health Organization, are "wastes having physical, chemical, or biological characteristics, which require special handling and disposal procedures to avoid health risks and/or other adverse environmental effects" (WHO, 1983). For example, toxic, explosive or flammable substances, or lead, mercury and phenols might become hazardous wastes. As virtually everything can become a waste and consequently a pollutant, only general guidelines on how to manage the waste problem can be given.

A waste policy at the enterprise level should comprise the following steps:

1. *Identify and control.* Know waste streams and classify them according to their harmfulness and their possibilities of reuse and recycling as the basis for further actions.

2. *Prevent and reduce.* Search for all possibilities to eliminate or reduce wastes at their source.

3. *Recycle and reuse.* Make best use of the existing wastes; use them for resource saving.

4. *Dispose.* Develop safe methods of disposal.

The criteria of occupational safety and health (see section 2.9) and preparedness for emergencies (see section 3.9) have to be applied to all the above steps. Let us now look in more detail at the four steps of waste management (see UNEP/IE/UNIDO, 1991).

1. Identification and control. Earlier in this book it was proposed that managers should draw in their minds a line around the enterprise and analyse which substances, materials and products were entering the plant and how they were leaving the plant.

The analysis of waste streams is parallel to the product flow and needs to be equally monitored. First, it is necessary to identify wastes associated with production processes. This can be done by a waste audit (see figure 24). Phases 1 to 4 comprise the analysis on which later action has to be based. There exist also rapid assessment methods which relate waste generation to the production processes responsible. These methods are especially useful when considering waste generation already at the planning phase of new projects.

Figure 24. Steps in carrying out a waste audit

Phase 1: Understand the process in the plant
List unit processes
Construct a process flow diagram

Phase 2: Define the process inputs
Determine resource usage
Check storage and handling losses
Record water usage
Determine current level of waste reuse

Phase 3: Define the process outputs
Quantify the process outputs
Account for wastewater flow and strength
Document wastes stored and disposed of

Phase 4: Carry out a material balance study
Summarize process inputs and outputs
Work out materials balance for unit processes
Evaluate the imbalance of materials
Refine the materials balance

Phase 5: Identify waste reduction options
List the obvious measures
Examine the problem waste streams
List the long-term options

Phase 6: Implement an action plan
Carry out a cost-benefit analysis for options
Select measures for immediate implantation
Start action on long-term measures

Source: Balkau, 1990, Annex 3 (adapted from: *Industrial waste audit and reduction manual,* Ontario Waste Management Corporation).

The rapid assessment methods compiled by the World Health Organization (WHO, 1982) into a manual of effluents, air emissions and solid wastes are a comprehensive set of tables from which waste loads generated by specific industry sectors and subsectors can be calculated. The input data are industrial output, in units appropriate to each industry, but usually by tonnes of product. The method is useful in order to obtain a first estimate of waste production from a single large company or group of companies whose production output is known (Balkau, 1990).

The identification of waste streams includes the classification of their harmfulness. There exist national and international lists of hazardous wastes such as, for example, the "orange book" concerning the transportation of dangerous goods. As waste streams become more and more international, they are increasingly being regulated by such international conventions as the Basel Convention on the Control of Transboundary Movements of Hazardous Wastes and their Disposal.

After knowing our waste streams and their nature we still have to decide how to handle them.

2. Prevention and reduction. Step 2 of a waste management strategy would be to prevent or reduce wastes, often called waste minimization. There are three main ways to eliminate or reduce pollution at the source:

☐ product modification or reformulation;

☐ process modification and maintenance; and

☐ equipment redesign.

Some companies have developed comprehensive PPP programmes in which research for waste elimination or reduction is systematically done. The box below shows the impressive results of the 3M Company's 3P programme for the 1975-89 period.

As has already been explained in section 2.3 on research and development management (pages 52–60), there is a great deal of scope for product

---

Pollution prevented and savings of 3M's 3P programme
The 3P record, 1975-89

Pollution prevented	United States	International	Total
Air pollutants (tons)	112 000	11 000	123 000
Water pollutants (tons)	15 300	1 100	16 400
Sludge/solid waste (tons)	397 000	12 000	409 000
Wastewater (billion gallons)	1	0.6	1.6
No. of approved 3P projects	785	1 726	2 511
Savings (US$ million)	426	74	500

Note: The figures represent savings only from the first year of each project. Projected over a period of several years, the pollution prevented becomes even more significant.

Source: 3M.

---

modification or, to use the terminology of the chemical industry, reformulations. The use of recyclable materials, product simplification by value analysis and careful thinking about packaging are amongst the most obvious solutions. Packaging, especially, constitutes a major problem area. It is not enough to use, for example, recyclable bottles; the whole concept of how to pack a product and how to re-collect the used bottles or other packaging materials has to be rethought. A German pharmaceutical producer, for example, is using so-called biopol bottles, which will be eaten up by microorganisms present in landfills.

Process changes are mostly associated with the installation of new equipment or equipment redesign. This might include more efficient burners in furnaces, as well as noise and heat insulation. Good housekeeping and maintenance are crucial in order to avoid leaks and spills, as well as to ensure that equipment performs within the given specifications. Many pollution problems in fact stem from older machinery or pipes which have not been well maintained.

3. Recycling and reuse. This method of dealing with waste within a closed production process can be an effective way to combine reduced consumption of resources with the prevention or elimination of pollution. Reuse of treated wastewater in a closed system, for example in paper manufacturing, and/or reuse of process heat by means of heat exchangers are typical examples.

Only where none of the following items – product modification, process change, equipment maintenance, equipment redesign, and/or in-process recycling – are possible or sufficient to meet environmental standards should end-of-the-pipe installations be considered. Management decisions should be taken based upon this order of priority.

Process change: Cost- and environment-effective

Painting with organic solvent-based paints poses not only the problem of how to dispose of the solvents, but also exposes workers to organic solvent vapours. Traditionally, pressures to reduce emissions have been met by costly end-of-pipe technology. A process change is less costly and, in addition, saves production costs as the example of a Swedish company shows. This company switched from organic solvent-based paints to powder paints. From the figures shown below, it becomes clear that not only are emission reduction and avoidance of workers' exposure to organic solvent vapours positive results of the process change, but that these also result in important economic benefits, which have been calculated by Backman et al. as follows:

Parameter	Costs (US$/yr.)
Paint (53,000 l/yr.)	322 200
Paint thinner (18,000 l/yr.)	22 500
Painting equipment cleaning (6 men, 1 day/month)	14 700
Hazardous waste management costs (combustion system)	34 000
Non-hazardous waste management costs	8 400
Powder paint (21,000 kg/yr.)	114 000
Four painters, full time	200 000
Combustion system fuel costs	40 000
Total annual costs before July 1989	755 800

Annual costs of the painting operations at Firm B before July 1989, at a time when most of the painting was being done with organic solvent-based paints.

Parameter	Costs (US$/yr.)
Powder paint (50,000 kg/yr.)	274 000
Two painters, full time	100 000
Total annual costs after July 1989	374 000

Annual costs of the painting operations at Firm B after July 1989, at a time when 95 per cent of the painting was being done with powder paints.

Thus, by changing over to powder paints, Firm B will save US$381,800 a year in operating costs. If one then adds the $34,000 for the savings from not having to pay the start-up costs of a combustion system, the first-year savings are $415,800. Consequently, the non-recurrent investment in the powder paint equipment of $383,000 will be paid back within the first year, and in addition there will be a net saving in the first year of $32,800. Subsequent annual savings will probably amount to more than $400,000, especially if calculations of decreased worker risks and decreased explosion risks are also included in the equation.

Source: Backman et al., 1990, p. 118.

Apart from in-process recycling and reuse there is the large field of recycling and reuse as an additional production step in the waste-generating company itself or another company. As disposal of wastes has become more and more expensive, recycling is an economic alternative and in many cases a profitable business. According to the business magazine *Fortune*, the recyclable material of choice these days is plastic.

A ton of recycled polyethylene terephthalate (PET), the plastic used to make soft-drinks bottles, can bring as much as US$1,000 – about the same as a ton of aluminium. Indeed, recycling plants are paying up to US$300 a ton for properly sorted plastic refuse dumped on their "tipping floors" (Sherman, 1989).

In Western industrialized countries approximately 25 to 40 per cent of the processed paper and 30 per cent of aluminium are recycled.

When considering the impact of recycling, the most significant aspect is the reduction in demand for the basic raw materials and energy, and the reduction in pollution from the manufacturing process. For instance, the recycling of a ton of aluminium, apart from the enormous saving in electrical energy, eliminates the need for 4 tons of bauxite and 700 kg of petroleum coke, whilst the emission of the air-polluting aluminium fluoride is reduced by some 35 kg. Paper is another case where the impact of recycling on the environment could be quite dramatic. Recycling paper not only spares millions of hectares of trees from felling, but also conserves energy and reduces water pollution (Kharbanda and Stallworthy, 1990).

A prerequisite for the reuse or recycling of wastes is their clear specifications and separate collection. Furthermore, proven and safe technologies should be used in recycling.

If recycling, such as plastics granulation or shredding, is not possible within the waste-generating company itself and there are no obvious buyers of the wastes, in some countries the waste disposal exchanges organized by industrial associations or chambers of commerce could help. To obtain reusable waste it might, however, be necessary to change the process. In a Scottish distillery, for example, which was threatened with closure because their still bottoms were polluting the local salmon river, they installed evaporators and driers. The dried "pollutant", which proved to be a very good animal feed, could then be sold.

4. Disposal. Even with increasing efforts of waste prevention, reduction and recycling there remain enormous amounts of wastes to be disposed of. The American Environmental Protection Agency proposes to incinerate wastes, where possible, in waste-to-energy plants. Landfills or deep well injections should, in the future, be reserved for those wastes which cannot be disposed of by other means.

Safe disposal, especially of hazardous wastes, implies a requirement to respect the relevant international agreements, especially those concerning transport, and occupational safety and health.

Table 12. Direct investment in energy conservation by Tata Iron and Steel Company (TISCO), 1981-87

Energy conservation areas	Total investment (Rs. million)	Cumulative savings (Rs. million)
Steam leaks, rationalization and insulation	5.50	6.60
Cold blast main insulation at blast furnaces	1.00	1.20
Insulation of LSHS storage tanks	0.15	0.12
Improvement in insulation in soaking pits; reheating furnaces and forge furnaces	2.00	1.80
Improvement in oil firing equipment and high efficiency burners in reheating furnaces and forge furnaces	2.00	3.00
Improvement in blast furnace stoves for higher hot-blast temperature	10.50	12.60
Soaking pit recuperator design changes	1.50	1.00
Improvement in furnace and energy distribution; instrumentation for energy conservation	38.00	45.60
L.D. gas recovery system and concast	140.00	390.10 (2 yrs.)
Utilization of waste L.P. nitrogen for steam savings	7.00	5.60 (4 yrs.)
Solar heating system for LSHS heating	0.45	0.41
Portable energy monitoring devices	1.50	1.35
Energy promotion activities, including interdepartmental competition	0.65	0.78 (est.)
Total	210.25	470.16

Source: Pachauri, 1990, p. 22.

Management strategy should follow the four steps explained in this section in order to deal effectively with wastes and pollution and to improve environmental performance in a cost-effective way.

3.8. Energy saving

Industry started to save energy, in some sectors as a major cost factor, long before people were made aware that burning fossil fuels contributes to the greenhouse effect. Even though most industries now use their energy much more efficiently than ten years ago, energy saving still constitutes a major cost-saving potential for enterprises.

It might be argued, especially by enterprises in developing countries, that increasing energy efficiency would require costly technology changes which they cannot afford, more especially if energy is still provided at a relatively cheap price. That investments in energy saving may well pay off,

Figure 25. Sources and transformation of energy

Resource		Conversion	Transmission and storage	Consumption
Renewable	*Non-renewable*	Burning (and steam generation)	Batteries	Lighting
Solar generation	Petroleum		Pipes	Processes
Water	Natural gas	Hydroelectric conversion	High voltage	Machine motion
Biomass	Coal		Transmission systems	
Wind	Nuclear materials	Electrolysis	Heat exchange	
Animal draught power		Photovoltaic conversion	Water tanks	
Geothermal energy		Wind electric conversion		
	(Firewood)			

however, can be demonstrated in the case of the Indian Tata Iron and Steel Company (TISCO) in the period 1981-87 (Pachauri, 1990). Table 12 shows energy conservation areas, total investments and the resulting savings.

Energy efficiency is not only the privilege of large companies, but it also has a major cost-reduction potential for small and medium-sized enterprises such as ceramics factories (IMR Corporation, 1983). Apart from industry there are immense energy-saving potentials in transport, agriculture and private households.

Where is energy lost or wasted and what strategy can be proposed by management to increase energy efficiency?

It is not only business, as the end user, which is concerned with increasing energy efficiency. We have to consider all stages from the selection of the energy resource, its conversion into heat, steam, electricity or mechanical power, transport and storage and finally the consumption for lighting, processes, machine motion, etc. (see figure 25). At each stage losses occur so that from the physical energy inherent in certain amounts of a natural resource only a small percentage reaches the user. Traditional combustion engines (e.g. generators) have an energy efficiency rate of 30 to 40 per cent, which means that 60 to 70 per cent of the fuel energy is "wasted". On the contrary, hydroelectric energy conversion reaches efficiency coefficients of over 90 per cent. If an enterprise has the choice, it should make use of renewable resources (e.g. install solar collectors for heating water) and select a conversion process with a high efficiency coefficient. Some enterprises, for example, have had positive experience with small hydroelectric power generators. Transmission and storage losses can be reduced by good insulation, well-manufactured batteries, regular cleaning of heat exchangers,

---

Improving the energy efficiency of major consumers

About half of all the energy used in the *chemical industry* is for steam and power and process heat, used principally for distillation, separation and pumping. Machine drive accounts for about 10 per cent, and electrolysis accounts for about 3 per cent. Opportunities for reducing energy use in chemicals production include upgrading electric motor efficiency, cogeneration, thermal recompression in evaporation and automated process control. These opportunities can be highly cost effective. An example of a process change with major energy efficiency benefits is the new Unipol process for making polyethylene, which uses only 35 per cent as much energy per pound of output as conventional processes. This process now accounts for 25 per cent of world production.

In cement making the inefficient wet process still accounts for the largest share of production, though the dry process uses one-third less energy per tonne. One hundred per cent penetration of the American market by the dry process, and associated efficiency improvement, could by 2010 reduce the energy required for cement making by at least 20 per cent in the United States alone.

In the paper industry promising technologies include continuous digesters, oxygen bleaching, upgraded evaporators, mechanical dewatering, boiler efficiency improvement, increased biomass use, and cogeneration. Paper drying has been estimated to be the single largest energy-consuming operation in the production of pulp and paperboard. Technical opportunities for efficiency improvement include: impulse drying, superheated steam drying with exhaust recompression, improved dewatering of sheet before drying and modern well-insulated drying enclosures (air recovery hoods).

Source: Chandler, 1990, p. 10.

---

and high-voltage transmission systems of electric energy, to mention only a few measures.

The main challenge for enterprises, however, is a reduction in overall consumption. Here are four main ways managers should consider when devising an improvement strategy:

☐ Questioning the need or justification for energy-consuming activities; for example: is the present level of lighting, heating and air-conditioning in our offices, stores or production facilities necessary? Can we reduce the number of our company cars and amount of miles driven (e.g. by switching from road to rail?).

☐ Avoiding energy losses;
for example: by better insulation of buildings, pipes and tanks.

☐ Improving the energy efficiency of processes;
for example: by substituting raw materials, selecting alternative sources of energy, changing the product and process technology. In lighting the use of low-energy bulbs can reduce the electricity bill by about 20 per cent.

☐ Reusing waste energy;
for example: by heat-exchange systems.

Energy questionnaire

Answers to the following can put potential energy savings into context

1. What proportion is your total energy cost as a percentage of
   (a) turnover;
   (b) manufacturing cost; and
   (c) profit?
2. Have the percentages changed over the past three years?
3. Did you achieve any energy savings in the past year? If so, how were the savings made? What was saved? What did it cost to make them?
4. What expenditure is planned for future energy purchases – say the next five years – and how will they affect your profitability?
5. Who monitors your organization's energy purchases, consumption and use?
6. Who is this person accountable to?
7. What is your present total energy cost?
8. What is the energy cost per unit of heat? (note: it helps to use consistent heat units throughout)
9. Do you know how much fuel is used in each
   (a) department; and
   (b) process unit (for example, furnace, boiler, vehicle)?
10. How are your total energy costs split between lighting, power, space heating, transportation?

Source: IPIECA and UNEP, 1991, p. 22.

In order to select those measures which with a minimum effort allow maximum results to be obtained, the biggest energy consumers in the enterprise have to be identified. It is advisable to carry out an energy audit and analyse the energy consumption of the cost centres. In applying a Pareto analysis the consumers should be ranked according to the energy consumed. After identifying the 20 per cent of internal consumers which are responsible for about 70 per cent of the energy costs, the above-mentioned four ways of reducing energy consumption should be pursued. Some companies make use of outside energy advisers or appoint an in-house energy adviser to run energy efficiency projects.

Energy consumption of equipment depends heavily on adequate maintenance. Therefore, maintenance units should be advised to include energy consumption in their maintenance checklists. Maintenance should also establish monitoring systems of energy consumption and act upon any perceived deviation from the established consumption standard. Energy saving, which combines cost reduction with environmental considerations, is a good starting-point for environmental management in a company. It is also a most appropriate field to initiate new approaches to employee participation.

3.9. Prevention of industrial disasters

Sixty thousand inhabitants were evacuated as a result of fire involving ammonium nitrate in France in October 1987. Fire involving methane caused four fatalities and one injury in Italy in April 1987. In Bulgaria a vinyl chloride explosion resulted in 17 deaths and 19 injuries in November 1986. An explosion involving fireworks killed 11 and injured eight in the Philippines in April 1986. In February 1986 a chlorine leak in the United States injured 76 persons. These cases are only a sample of reported events.[1]

Even more disastrous events may be cited. These include the release of the chemical methyl isocyanate in Bhopal, India, in 1984, resulting in over 2,000 fatalities and 200,000 injuries. Two weeks earlier there was an explosion of liquefied petroleum gas (LPG) in Mexico City, resulting in 650 deaths and several thousand injuries. In 1976, 30 people were injured and 220,000 persons were evacuated from surrounding villages when a process malfunction resulted in a small release of the chemical dioxin in Seveso, Italy. The human and economic damage resulting from all these events and many others is huge.

Although these cases may differ in the way they happened and the chemicals that were involved, they share a common feature: they were uncontrolled events involving fires, explosions or releases of toxic substances that either resulted in the death and injury of a large number of people inside and outside the plant or caused extensive property and environmental damage, or both. The storage and use of flammable, explosive or toxic chemicals having the potential to cause such disasters are generally referred to as major hazards. This potential, therefore, is a function both of the inherent nature of the chemical and the quantity that is present on site.

There is a great variety in the types of major accident that can take place, leading to the concept of a "major hazard" as an industrial activity requiring controls over and above those applied in normal factory operations, in order to protect both workers and people living and working outside. These controls form an integrated package – a major hazard control system – which aims not only at preventing accidents but also, and most importantly, at mitigating the consequences of any accidents which do take place.

[1] This section is based on ILO, 1988c, and UNEP, 1988.

Figure 26. EU Directive criteria for major hazard installations

---

Toxic substances (very toxic and toxic)

Substances showing values of acute toxicity and having physical and chemical properties capable of entailing major accident hazards (for details see ILO, 1988c).

Flammable substances

1. Flammable gases: substances which in the gaseous state at normal pressure and mixed with air become flammable and the boiling-point of which at normal pressure is 20 °C or below.

2. Highly flammable liquids: substances which have a flash-point lower than 21 °C and the boiling-point of which at normal pressure is above 20 °C.

3. Flammable liquids: substances which have a flash-point lower than 55 °C and which remain liquid under pressure where particular processing conditions, such as high pressure and high temperature, may create major accident hazards.

Explosive substances

Substances which may explode under the effect of flame or which are more sensitive to shocks or friction than dinitrobenzene.

Source: ILO, 1988c, p. 8.

Because of the complexity of the industrial activities concerned, major hazard control needs to be based on a systematic approach which is described in detail in the ILO manual *Major hazard control,* from which this section has been adapted (ILO, 1988c). The manual is a useful guide for managers who want to devise a major hazard control system for their enterprise. The ILO code of practice *Prevention of major industrial accidents* (ILO, 1991b) also provides valuable guidance.

The basic components of this system of which managers should be aware are:

(a) *Identification of major hazard installations*

It is necessary to identify the installations which, according to the definition, may fall within the criteria set for the classification of major hazard installations. Government authorities and management should institute the identification of major hazard installations on a priority basis. Identification may be done in accordance with national and international guidelines. The European Union, for example, has adopted a Council Directive on major accidents and hazards of certain industrial activities (82/501/EEC) which sets criteria to identify major hazard installations. Relevant criteria and priority chemicals are given in figure 26 and table 13.

(b) *Information about the installations and assessment of major hazards*

Once the major hazard installations have been identified, further information needs to be collected about their design and operation. In addition, such information must also describe all other hazards specific to the

Table 13. Priority chemicals used in identifying major hazard installations

Name of substance	Quantity > (t)
General flammable substances	
Flammable gases	200
Highly flammable substances	50 000
Specific flammable substances	
Hydrogen	50
Ethylene oxide	50
Specific explosives	
Ammonium nitrate	2 500
Nitroglycerine	10
Trinitrotoluene	50
Specific toxic substances	
Acrylonitrite	200
Ammonia	500
Chlorine	25
Sulphur dioxide	250
Hydrogen sulphide	50
Hydrogen cyanide	20
Carbon disulphide	200
Hydrogen fluoride	50
Hydrogen chloride	250
Sulphur trioxide	75
Specific very toxic substances	
Methyl isocyanate	150
Phosgene	750

Source: ILO, 1988c, p. 8.

installation. Because of the likely complexity of the installation, the information should be gathered and arranged systematically, and should be accessible to all parties concerned within the industry, such as management and workers, and outside the industry, such as the government bodies which may require it for licensing and inspection purposes. In order to achieve a complete description of the hazards, it may be necessary to carry out safety studies and hazard assessments to discover possible process failures and to set priorities during the process of hazard assessment. Rapid ranking methods may be used to select the units which require a more thorough assessment.

(c) *Control of the causes of major industrial accidents and safe operation*

In addition to preparing a report, management has the primary responsibility of operating and maintaining a safe plant. A sound safety policy is required. Technical inspection, maintenance, plant modification, training and selection of suitable personnel must be carried out according to sound procedures. In addition to the preparation of the safety report, accidents should be investigated and reports submitted to the authorities. Lessons should be learned.

Assessment of the hazards for the purposes of licensing, where appropriate, inspection and enforcement of legislation are the responsibility of government in the control of major hazards. Land-use planning can appreciably reduce the potential for a disaster and will probably come under government control. The training of factory inspectors, including chemical inspectors, is also an important government role.

Responding effectively to emergencies

☐ There are emergency plans covering all the significant on-site risks which are available to all relevant personnel.

☐ The emergency plans:
- clearly define responsibilities;
- clearly define all procedures;
- define environmental sensitivities;
- define linkages to local authorities and other bodies;
- define linkages to other plans or resources; and
- are regularly updated.

☐ Key personnel are familiar with the provisions of the plan.

☐ Equipment and manpower of adequate quantity and quality are available for emergency response:
- from on-site resources; and
- from identified local and national resources.

☐ Response equipment is well maintained and accessible.

☐ There is an established programme for the training of personnel in the use and handling of emergency response equipment.

☐ There is an established programme of emergency plan exercises, with systems for implementing resultant recommendations.

☐ Incident review procedures are formalized and recommendations are implemented.

☐ All incidents are recorded and reported, as appropriate, to management and regulatory authorities.

Source: Adapted from UNEP/IEO, 1990b.

Figure 27. The responsibility bridge of APELL

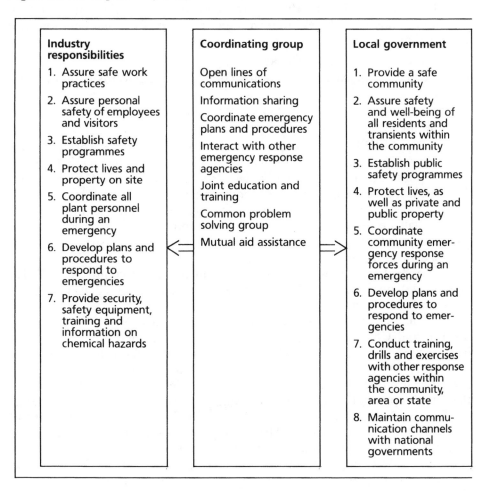

Industry responsibilities	Coordinating group	Local government
1. Assure safe work practices	Open lines of communications	1. Provide a safe community
2. Assure personal safety of employees and visitors	Information sharing	2. Assure safety and well-being of all residents and transients within the community
3. Establish safety programmes	Coordinate emergency plans and procedures	
4. Protect lives and property on site	Interact with other emergency response agencies	3. Establish public safety programmes
5. Coordinate all plant personnel during an emergency	Joint education and training	4. Protect lives, as well as private and public property
6. Develop plans and procedures to respond to emergencies	Common problem solving group / Mutual aid assistance	5. Coordinate community emergency response forces during an emergency
7. Provide security, safety equipment, training and information on chemical hazards		6. Develop plans and procedures to respond to emergencies
		7. Conduct training, drills and exercises with other response agencies within the community, area or state
		8. Maintain communication channels with national governments

Source: UNEP/IEO, 1988, p. 22. See Appendix 3, "Useful addresses".

(d) *Emergency planning*

Previous sections focus on the prevention of major accidents. However, while all accidents can in principle be prevented, some accidents will still happen and it is essential to be well prepared for them. Loss of life, injury and damage to the environment and property can all be avoided or at least minimized if people know what to do in an emergency and correct first response measures are taken.

The consequences of major accidents do not necessarily stop at the factory fence. It is vital (literally) that on-site and off-site planning be properly integrated.

Awareness and Preparedness for Emergencies at Local Level (APELL) was launched in 1988 by the United Nations Environment Programme's Industry and Environment office (UNEP IE), in cooperation with governments and industry. Its objectives are to:
- encourage prevention of accidents
- create and/or increase public awareness of possible hazards within a community
- stimulate the development of cooperative plans to respond to any emergency which might occur.

The APELL Handbook, now available in 20 languages worldwide, sets out a ten-step process to guide local communities in strengthening their emergency response capability. Its basis is the APELL Coordinating Group, including representatives of industry, local government and the community (see figure 27 on page 151). The Group's responsibility is to assure the production of a coordinated emergency plan and encourage community awareness and preparedness, not to take charge of an emergency – the command structure for this will be defined in the plan.

APELL has now been introduced in about 30 countries. "APELL Worldwide", published by UNEP IE, gives national overviews of its implementation in 12 of them.

The APELL Newsletter appears twice a year in English, French, Spanish and Chinese. The United States Environmental Protection Agency has adapted its Computer-Aided Management of Emergency Operations (CAMEO) software for international APELL applications. UNEP IE has published a number of technical guides to help implement APELL, some of them in cooperation with other international organizations including IMO, IPCS, OECD and WHO. "Safety, Health and Environmental Management Systems" and "TransAPELL" (APELL applied to accidents arising from the transport of dangerous goods) are in preparation.

The APELL Programme Coordinator at the UNEP Industry and Environment office may be contacted by those who would like to participate in the APELL programme (see Appendix 3, "Useful addresses", p. 183).

The prevention of industrial disasters cannot only rely on adequate emergency plans and other organizational precautions but requires a management attitude to assume full responsibility for a company's activities.

(e) *Inspection of major hazard installations*

Major hazard installations should be regularly inspected in order to ensure that the installations are operated according to the appropriate level of safety (ILO, 1991b). This inspection should be carried out both by a safety team which includes workers and workers' representatives and separately by inspectors from competent authorities. Both types of inspection may be carried out in other ways where appropriate. Safety personnel from the installation within this safety team should be independent of production line management and should have direct access to works management.

4

Institutional support

4.1. Employers' organizations and business associations

Up to now this book has dealt with solutions for individual enterprises on how to incorporate environmental management into their corporate activities.

How to achieve profitability, growth and survival, produce goods and services demanded by consumers, and provide employment and satisfactory working conditions for their workers, while at the same time refraining from damaging the physical environment through pollution and the wasteful use of scarce resources, is not only a matter for individual enterprises but also for the business community as a whole (see Evan-Stein, 1990).

There are many reasons for enterprises to undertake collaborative efforts in environmental matters. Firstly, there is an enormous need for information on "how to become Green and clean". Secondly, enterprises might feel threatened by mounting public pressure and ever tougher government legislation. Thirdly, enterprises would like to make sure that the general public is made aware of their precautions to protect the environment. By collaborative efforts enterprises will have easier access to the media to persuade the public that industry is genuinely concerned with environmental protection.

A fourth reason for collaborative efforts concerns maintaining a constructive dialogue with government regulatory agencies about environment-related regulations. The purpose of this is to ensure, as far as possible, that any new regulations that come into force are realistic, both technically and economically. It is not to prevent the development of new regulations. The Green movement is nowadays too firmly established for such a strategy to have any hope of long-term success.

At the local, sectoral, regional, national and international levels employers and business leaders cooperate in employers' organizations, industrial associations, chambers of commerce, business round tables, and environmental management associations. Industrial associations have become quite active in environmental matters. The following may be cited: the International Petroleum Industry Environmental Conservation Association (IPIECA), the

European Council of Chemical Manufacturers' Federations (CEFIC) and the International Group of National Associates of Manufacturers of Agrochemical products, all belonging to "pollution prone" sectors.

The International Chamber of Commerce (ICC), which issues periodically updated environmental guidelines for world industry, has set up a specialized International Environmental Bureau which provides information and clearing-house facilities, with the support of some leading transnational corporations.

The World Business Council for Sustainable Development (WBCSD) has become widely known through its report *Changing course*, which was prepared under the leadership of the Swiss industrialist Stefan Schmidheiny for the Rio Conference. It is now undertaking a number of initiatives to promote the concept of eco-efficiency, and to study the relationships between trade and the environment, among others.

The Employers' Activities Bureau of the International Labour Organization has an extensive environmental training programme for employers' organizations in developing countries and provides institutional assistance to improve the environmental capabilities of these organizations (Evan-Stein, 1990).

Employers' organizations in some countries have also established contacts with trade unions regarding environmental matters. In his 1990 Report to the International Labour Conference entitled *Environment and the world of work*, the Director-General of the ILO stated that, in order to implement effectively the various environmental policies being developed by enterprises, employers' organizations and industrial associations, the close collaboration of well-trained and environmentally aware employees is essential and may provide new opportunities for closer collaboration on the environment between employers and workers in the future (ILO, 1990).

What could be the task of an employers' organization in the field of the environment? In a number of regional meetings held within the above-mentioned ILO environmental training programme for employers' organizations, it was concluded that they may be able to play a useful role mainly in the following areas:

Environmental legislation. They should seek to be consulted on a regular basis in the drafting of environmental laws, regulations and standards, as well as the modalities of their application.

Environmentally sound technologies. They should provide a clearing-house service for their members concerning environmental technologies, centralizing access to international information networks.

Environmental awareness. They should promote environmental awareness among their members through information, education and training.

Training. They should integrate general environmental training, especially for managers, within their traditional training activities.

Exchange of information. They should facilitate the exchange of relevant information both between members and with other employers' organizations, regionally and internationally.

In a number of countries employers' organizations have established an environment focal point at their secretariat. Other organizations have opted to include the environment in the functions of their occupational safety and health desk. More detailed information about employers' organizations and the environmental challenge can be found in the training material which supports the ILO environmental management training course for employers' organizations (Evan-Stein, 1990).

In some countries organizations of young entrepreneurs have undertaken to promote environmental issues. The respective German organization, for example, has published a collection of checklists on environmental protection as part of the enterprise strategy (BJU, 1990).

Environmental management associations or business circles were created in a number of countries by environmentally minded business leaders. Such associations organize congresses, seminars and training courses, assist projects, promote research, disseminate information and collaborate in contributions to mutual and international networks.

The participation of enterprises in one or the other of these collaborative organizations not only makes the voice of an enterprise heard, but can also demonstrate a sense of social responsibility.

Selected addresses of business organizations dealing with the environment can be found in Appendix 3.

4.2. Workers' organizations

The environmental movement has proved to be a real challenge for trade unions and workers' representatives.

In addition to all their other roles and responsibilities, workers and their organizations are also "environment and development" organizations by virtue of the extremely high priority they have given to the protection of the working environment and the promotion of equitable social and economic development. Trade unions are increasingly aware that employment and income opportunities, the working environment and the natural environment are all closely interlinked. Furthermore, the extensive membership and the existing network of contacts and collaboration among trade unions provide important channels through which the concepts and practices of sustainable development can be supported. Their role, and particularly the capacity to collaborate with other social partners (e.g. governments, industry and NGOs), will need to be strengthened in order to facilitate the achievement of sustainable development.

Management should make full use of the knowledge, experience, commitment and influence of their workers, and trade unions and the community, to make sustainable development a common cause.

In his 1990 Report the Director-General of the ILO listed possible ways of worker involvement (ILO, 1990). Activities and potential roles for workers and their organizations, at the enterprise level, to improve environmental performance include the following.

☐ Given their day-to-day experience in the workplace, workers and their organizations can make an important contribution to the improvement of the working and general environment and should make a point of ensuring that they are informed and consulted at an early stage on all environment questions.

☐ The active involvement of workers and their organizations should be sought concerning the design and implementation of all environmental policies or programmes which might promote new employment, protect existing employment, or lead to a loss of employment (with the consequent need to institute adequate "safety-net" measures).

☐ Workers need access to information on – if not actual participation in – the establishment of company environmental strategies or policies; the introduction of new technologies; monitoring of chemical emissions inside and outside the plant; environmental audits; inspections and reports regarding compliance with environmental regulations and standards, etc.; in other words, they need "the right to know".

☐ Workers should participate in the design and development of training programmes for workers and management to provide environmental awareness and the skills necessary to meet environmental objectives; special efforts should be made to ensure that workers' safety and health representatives or special environmental representatives receive appropriate environmental training.

☐ Workers should promote the setting up of special joint committees to deal with general environment issues or the broadening of the mandate of existing joint committees (e.g. safety and health committees) to encompass the general environment.

☐ Workers should consider the creation of special trade union environmental committees, if appropriate.

☐ Workers should actively participate in environmental activities within the local community, and facilitate exchanges of views on potential problems and activities of common concern.

☐ Workers may be able to influence the purchasing policies of enterprises, consumers and government in order to identify and promote products which are safe for the working and general environment, by such means as the establishment of standards on environmental certification or labelling.

☐ Workers should urge employers to recognize good environmental performance of workers and management and to provide incentives for such performance.

☐ Workers should cooperate with appropriate government inspection authorities to ensure the improvement and enforcement of regulations and standards on the working and general environment.

☐ Workers should collaborate with employers, governments and consumers to ensure that the "polluter pays principle" is enforced and that the costs are not simply passed on to consumers without the polluter making investments to prevent further pollution.

With a change in industry structure due to environmental legislation and consumer pressure, the composition of the workforce will also undergo changes. This will have repercussions on trade union structure and affiliation. As trade unions are a powerful force in society, the unions are expected to express their views on sustainable development as well as on specific environmental issues. For many workers' organizations a starting-point to deal with broader environmental issues was the working environment. The relation between working environment and outside environment from a trade union point of view is well reflected by the words of Mike Wright, Director of Safety and Health, United Steel Workers of America (ICFTU, 1990):

⋯ We've learned through bitter experience that there is no fundamental difference between the working environment and the environment outside the plant, between the way a company treats its workers and the way it treats its neighbours. You can't ban leaks and spills in the company rule book. A company has to act responsibly towards both its employees and the community, or it won't do either ⋯

The International Confederation of Free Trade Unions (ICFTU) is particularly active in environmental matters. It has established a working party on occupational safety and health and the working environment, which meets once a year and brings together trade unions specifically with the aim of exchanging views and experiences, as well as trying to identify future action priorities.

Many national trade union centres, their affiliates and international secretariats have also established occupational safety and health and environment (OSHE) working parties and participate in tripartite committees to improve the working and general environment.

The ILO is carrying out a workers' education project to assist union leaders and workers' education officers to integrate environmental considerations into their work. The project aims at raising awareness among trade unionists about the importance of environmental matters and to devise union strategies to promote environmental protection.

As both employers and workers have a stake in the environment, this is also an excellent field for collaboration. Joint statements of environmental policy may be crucial. Such a statement has been made, for example, by the Federation of Netherlands Industry (VNO) and the Federation of Netherlands Trade Unions (FNV). The field of energy conservation also calls for joint programmes between employers and workers. In such collaboration the government could also participate, thus giving programmes a tripartite character (see ILO, 1990).

4.3. Environmentalist groups

Green pressure groups are still very much feared by industrial managers. It is a nightmare for a plant manager to imagine that the main factory entrance might be blocked by Green activists and the plant receive extensive press coverage. How should the plant manager react?

Environmentalist groups have been very creative in brandmarking the environmental damage caused by enterprises. Green pressure groups have woven a complex net of lobbying relations. They influence legislation and act as watchdogs to enforce environmental regulations. Green groups are represented in local, regional, national, and international parliaments and political forums. Green activists have become ministers and high officials charged with environmental protection. Industry has had to learn that it might well be on the losing side if it enters into confrontations with these Green pressure groups.

This is one reason why many enterprises have started to adopt a collaborative rather than a confrontational strategy (see section 3.5). Green activists are invited to discussions with managers, or are asked for their advice on how a company could develop a Green profile. The change from confrontation to collaboration is difficult not only for industry but also for the pressure groups, who fear the loss of support from their more critical members. Pressure groups also have to develop dialogue skills.

Managements deciding to reach out their hand to Green activists will need a lot of patience in getting a dialogue started, and should not expect to "brainwash" environmentalists. On the other hand, Green pressure directly or indirectly via managers' families or local environmental awareness-raising has opened the eyes of a significant number of managers. They now perceive that the run to maximize turnover and return on investment has blinded them to the destructive consequences of their actions.

In section 3.5 we discussed communication on environmental matters, but with whom could managers communicate if they decided to do so?

Indeed, the number and orientations of environmentalist groups are so manifold that it might be difficult for a manager to find the right discussion partners. There are small local groups, regional and national movements, and there are the multinationals such as Greenpeace, Friends of the Earth and the World Wide Fund for Nature (WWF). Their members are counted by hundred thousands and some of the pressure groups are also quite strong economically. If we take, for example, Friends of the Earth, one of the leading environmental pressure groups in the United Kingdom, they have a network of more than 270 local groups across the country, which are able to threaten or organize country-wide boycotts against, for example, ozone-damaging aerosols.

These lobbyists work with all political parties, commission research, and coordinate Earth action groups of young people. Moreover, Friends of the Earth is represented in 35 countries, including developing countries.

Greenpeace, which has almost 400,000 members in Britain alone and 2 million in the United States, is known for its spectacular actions such as blocking ships transporting hazardous waste or climbing plant chimneys. Greenpeace also has a reputation for running its business effectively.

The World Wide Fund for Nature states in information brochures that it has more than 4 million supporters worldwide, and 28 official and associate national organizations on five continents.

In the United States the National Wildlife Federation (NWF), which was founded in 1936 to protect wildlife, has become a major Green pressure group dealing with issues such as safe drinking water, clean air, and

---

Some rules on how to deal with environmental groups

1. Accept environmentalist groups as critical challengers of your business activities.
2. Enter into an open dialogue with environmentalist groups without pretending to be Greener than your company is.
3. Avoid any deals with environmentalist groups which can be seen as "bribes to keep quiet".
4. Be aware of conflicting views within the Green movement.
5. Observe the development of the Green agenda.
6. Establish contacts and support activities of the community as a whole.
7. Seek a multilateral dialogue (with trade unions, citizen action groups, local authorities, political parties, churches, etc.).
8. Act together with the business community (via business associations, employers' organizations, chambers of commerce, etc.).

---

environmentally harmful industrial products. With 5.8 million members and a budget of US$85 million there is sufficient power to influence public opinion, legislation and business.

Apart from campaigning, Green pressure groups have developed a number of activities and services. Some environmentalist groups have started consultancy activities. An American environmentalist group, for example, reached an agreement with a fast-food chain to examine ways to reduce the amount of waste created by this kind of business. As part of the deal the fast-food chain was not allowed to mention the assistance of the environmentalist group in any marketing activity while the environmentalists were free to publish the findings (*The Economist*, 20 Oct. 1990).

Some environmentalist groups support environmental research institutes. Perhaps the most famous of these institutes is the World Watch Institute in Washington, DC, which in collaboration with the UNEP annually issues an informative and sometimes controversial report on the state of the world (World Resources Institute, 1990). Other institutes specialize in alternative materials and processes. R&D managers, especially, should be interested in making use of their experiences.

There are also ecologically committed institutes which carry out consumer product testing, and produce specialized journals or handbooks. Since they are able to significantly influence consumers, marketing managers should establish contact with these institutes.

While environmental groups first emerged in industrialized countries, they are now also fairly well established in developing countries. They often link environmental issues to development and social justice problems in these countries, for example, as in the case of the protection of tropical rain forests. The power of some Green movements in developing countries can be demonstrated by the fact that some of their leaders have been charged by governments with environmental protection.

Green pressure groups would not be as effective as they are without the support of the media. Newspapers, radio and television, journals, books and information leaflets now cover environmental matters regularly. They thus form a Green pressure group in their own way. Managers should seek active contact with the media and have an open house for them. It is much more effective to contribute positively to environmental reports than to quarrel later about the contents of a reportage, or place expensive disclaimers, which try to convince critics that enterprise X is not a major polluter as the local media and Green activists are claiming. However, there is, of course, a precondition to this open-house policy: the enterprise should have nothing to hide.

To facilitate the identification of and contact with some of these major environmentalist groups and institutes, their addresses have been listed in Appendix 3.

4.4. Government institutions and international organizations

Compliance with environmental legislation is not always an easy task for enterprises. Knowing which government department might be in charge of environmental matters can be less than obvious. Depending on the country and the subject-matter, the following might be involved: the Environmental Protection Agency; Labour Inspection; Ministry of Commerce in cases concerning export or import licences; National Standardization Organizations; National or Regional Planning Commissions or Local Authorities, to cite just some possibilities. For managers there are no general guidelines on how to find their way through this jungle. In countries where a well-established Environmental Protection Agency (EPA) exists this would perhaps be the most effective contact partner. Such EPAs often have extensive established information services. They are the national focal points of the INFOTERRA database of the United Nations Environment Programme (UNEP), which provides most of the environment-related data available.

The role, size and structure of environmental agencies vary from country to country. In the United States, for example, the EPA is a regulatory agency whose main job is to develop, monitor and enforce laws delegated to it by Congress. The Agency administers laws that cover a variety of pollution prevention and abatement issues, such as air pollution, drinking-water, pesticides, toxic chemicals and waste disposal.

Each law provides a framework within which the EPA has to prepare detailed regulations. These functions are supported by a wide range of activities, including extensive scientific research and public education. The EPA is not chiefly responsible for many other environmental issues, such as national parks and forests (US Department of the Interior), oceans (US National Oceanographic and Atmospheric Administration) or energy (US Department of Energy). Altogether, the EPA has about 16,000 employees. They are distributed among its headquarters in Washington, DC, ten regional offices and numerous research facilities (Halther, 1990).

Apart from EPAs, the regional and local planning authorities are the agencies with major responsibility for the extension of existing, or the building of new, production facilities. These authorities are especially exposed to the NIMBY syndrome. Many countries have seen cases where new projects which had already been approved by these authorities have been stopped by the courts based on appeals by environmentalist groups. These authorities, and the enterprises which ultimately pay the bill when project approval is not granted, have had to learn that the community has to be integrated into the planning process.

In overseeing a company's operations there are again a number of government organizations involved. While in some countries partial or complete responsibility lies with the EPA, in other countries environment has been added as an additional task of labour inspection.

Notwithstanding the allocation of responsibilities, most countries have problems with effectively enforcing their environmental regulations as they lack qualified staff who can keep pace with technological developments.

This is particularly true in the case of developing countries and it is in this area where international organizations can offer the most assistance.

International organizations

The *United Nations Environment Programme (UNEP)* has created its *Industry and Environment Office (IEO)* in Paris and offers a number of useful services to industry. It publishes technical reviews and guidelines on clean technologies, issues a quarterly review, entitled *Industry and Environment*, and runs a query-response service. On government request, diagnoses of environmental problems are carried out and practical solutions are studied. Training and awareness-raising is a further activity of UNEP/IE. In section 3.9 the *APELL programme* of awareness of and preparedness for emergencies was mentioned as another service. UNEP/IE in cooperation with the US Environmental Protection Agency also runs the *International Cleaner Production Information Clearing House (ICPIC)*. ICPIC contains a computerized information system on clean technology in its widest sense. The system is available to anyone with access to a computer and modem. There is no user fee.

The OzonAction Programme to enable developing countries and countries in transition to implement the Montreal Protocol on Substances that Deplete the Ozone Layer has been mandated to be an Information Clearinghouse under the Multilateral Ozone Fund. It publishes technical and policy documents and provides "train the trainer" training courses for good-practices in refrigeration, safety and monitoring of ozone-depleting substances. It provides information on diskettes (OzonAction Information Clearinghouse – Diskette Version) to developing countries and runs networks of Ozone Officers in Africa, Latin America and South-East Asia and the Pacific.

Chemical safety is another field where international cooperation and assistance is needed.

In this regard, the *UNEP's International Register of Potentially Toxic Chemicals (IRPTC)* and the *ILO's Occupational Safety and Health Branch* work in close collaboration, especially in collecting and compiling information on legal restrictions or the use of chemicals at work. The *UNEP/ILO/WHO International Programme on Chemical Safety (IPCS)* has produced a set of environmental health criteria documents dealing with chemical and physical pollutants. The IPCS is devising a system of International Chemical Safety Guides and International Chemical Safety Cards to provide specific information for the users of chemicals.

The *International Occupational Safety and Health Information Centre (CIS)* is the ILO's focal point for the UNEP's network of national environment-information centres (INFOTERRA) and makes active use of the latter's facilities to disseminate information on chemicals. The CIS further collaborates with over 100 national CIS centres (approximately half of which are in developing countries) with respect to the collection, processing and dissemination of information as a means of improving the working and general environment. The CIS database (CISDOC) includes more than 4,000 chemical safety data sheets in several languages and pointers to tens of thousands more. Each sheet gives detailed advice on chemical hazards and on precautionary measures, including those related to waste removal and disposal.

Further ILO assistance for workers' and employers' organizations has already been mentioned in the previous sections. In addition, training materials for managers have been developed by the Entrepreneurship and Management Development Branch.

UNESCO's International Environmental Education Programme (IEEP) deals with the exchange of information on environmental education activities, a monthly *CONNECT* newsletter is issued and an International Directory of Educational and Training Institutes provides details of environment-related programmes and courses. Pilot projects, curriculum development and trainer's training form an integral part of the programme.

UNIDO assists industry in developing countries to introduce clean technologies.

The *World Health Organization (WHO)* and the *Food and Agriculture Organization (FAO)* are other United Nations organizations which can assist industry in becoming clean and Green.

Useful information for enterprises is also contained in a number of studies published by the Organisation for Economic Co-operation and Development (OECD).

The message of many of these studies is that there are economically feasible ways of becoming an "ecological" industry. One of the ways, the tripartite collaboration of government, employers and workers, is an appropriate mechanism to bring sustainable development a step closer.

Appendices

1. Bibliography

Backman, M., et al. 1990. "Preventive environmental protection strategy: First results of a Swedish experiment in Landskrona", in International Environmental Bureau (ed.): *Special wastes*. Geneva, pp. 108-126.

Baier, P. 1969. *Wertgestaltung*. Munich, Hanser.

Bakar Che Man, A.; Gold, D. 1993. *Safety and health in the use of chemicals: A training manual*. Geneva, ILO.

Balkau, F. 1990. *Pollution control and low waste technologies in agro-based industries*. Paris, UNEP/IEO.

Baumert, K.; Hermann, M. 1996. "Successful environmental reporting by target group orientation", in North (ed.), 1996.

BJU (Bundesverband Junger Unternehmer). 1990. *Umweltschutz als Teil der Unternehmenstrategie*. Bonn.

Blohm, H., et al. 1982. *Produktionswirtschaft*. Berlin, Verlag Neue Wirtschafts-Briefe.

Boland, R.G.A. (ed.). 1986. *Environmental management training* (Vols. 1 to 5). Geneva, ILO/UNEP.

Brackley, P. 1988. *Energy and environmental terms: A glossary*. Aldershot, Hampshire, Gower.

Bringer, R.P.; Benforado, D. 1989. "Pollution prevention as corporate policy: A look at the 3M experience", in *Environmental Professional*, 11 (1989), pp. 117-126.

Cairncross, F. 1991. *Costing the earth*. London, Economist Books.

Centre de documentation sur les déchets. 1989. *Clean technologies?* (educational pamphlet). Neuilly-sur-Seine.

CEP (Council on Economic Priorities). 1989. *Shopping for a better world: A guide to socially responsible supermarket shopping*. New York.

Chandler, W.D. 1990. "Industrial energy efficiency: A policy perspective", in *Industry and Environment*, Vol. 13 (1990), 1, pp. 9-12.

Commission of the European Communities. 1990. *Employment in Europe*. Brussels, COM (90) 290 Final.

ECOTEC. 1990. *The impact of environmental management on skills and jobs in Birmingham*. Birmingham, ECOTEC Research and Consultancy Ltd.

Ehrlich, P.R.; Ehrlich, H.H. 1991. *Healing the planet*. Reading, Massachusetts, Addison-Wesley.

EIASM (European Institute for Advanced Studies in Management). 1991. "The greening of accounting", in *EIASM Newsletter*, Spring 1991, p. 1.

Elkington, J.; Burke, T. 1989. *The green capitalists*. London, Gollancz.

Elkington, J.; Hailes, J. 1989a. *The green consumer guide*. London, Gollancz.

—; —. 1989b. *Universal green office guide*. London, Universal office supplies.

Elkington, J., et al. 1990. *The green consumer: A guide for the environmentally aware*. Harmondsworth, Middlesex, Penguin.

—. 1991. *The green business guide*. London, Gollancz.

Evan-Stein, H. 1990. *Environmental management training course for employers' organisations (Environmental law and administration; The human environment: its functions and uses; Environmental planning and management; Pollution prevention pays; Employers' organisations and the environmental challenge; Instructors guide)*. Geneva, ILO/UNEP; Employers' activities doc. EMP/ENV.

Fritz, W. 1995. "Umweltschutz und Unternehmenserfolg" [Environmental protection and business success], in *DBB*, Vol. 55, No. 1, pp. 347–357.

Gladwin, T.N. 1980. "Patterns of environmental conflict over industrial facilities in the United States, 1970-78", in *Natural Resources Journal*, 20 Apr. 1980, p. 249.

Halther, F. 1990. "How to obtain information from the US Environmental Protection Agency", in *UNEP Industry and Environment*, Jan.-Feb.-Mar. 1990, pp. 33-35.

Henning, n.d. *Environmental affairs glossary*. Billings, Montana, Eastern Montana College.

—. 1989. *Environmental auditing*. Paris.

ICC (International Chamber of Commerce). 1990. *ICC position paper on environmental labelling schemes*. Paris.

ICFTU (International Confederation of Free Trade Unions). 1990. *Occupational health, safety, and the environment in Central and Eastern Europe*. Brussels.

IILS (International Institute for Labour Studies). 1991. *Environment – employment – new industrial societies: A bibliographic map* (Série bibliographique). Geneva.

ILO (International Labour Office). Forthcoming. *Encyclopaedia of occupational health and safety*. Geneva, 4th ed.

—. 1984. *Protection of workers against noise and vibration in the working environment*. An ILO code of practice. Geneva.

—. 1985. *Occupational exposure to airborne substances harmful to health*. An ILO code of practice. Geneva.

—. 1987. *Radiation protection of workers (ionising radiations)*. An ILO code of practice. Geneva.

—. 1988a. *Higher productivity and a better place to work. Trainers' manual* (by J.E. Thurman, A.E. Louzine and K. Kogi). Geneva.

—. 1988b. *Safety in the use of asbestos*. An ILO code of practice. Geneva.

—. 1988c. *Major hazard control. A practical manual.* Geneva.

—. 1989. *Employment and training implications of environmental policies in Europe,* Geneva, Tripartite Meeting of Experts on Employment and Training Implications of Environmental Policies in Europe, Geneva, 1989 (doc. ETIEPE/1989/1).

—. 1990. *Environment and the world of work.* Report of the Director-General (Part I), International Labour Conference, 77th Session, Geneva, 1990.

—. 1991a. *Safety and health in the use of agrochemicals. A guide.* Geneva.

—. 1991b. *Prevention of major industrial accidents.* An ILO code of practice. Geneva.

—. 1993. *Safety in the use of chemicals at work.* An ILO code of practice. Geneva.

—. 1995. *Safety, health and welfare on construction sites. A training manual.* Geneva

—. 1996a. *Ergonomic checkpoints. Practical and easy-to-implement solutions for improving safety, health and working conditions.* Geneva.

—. 1996b. *Recording and notification of occupational accidents and diseases.* An ILO code of practice. 1996.

IMR Corporation (ed.). 1983. *Reducing energy costs in small business.* Reston, Virginia, Reston Publishers Company.

International Management. 1990. Autumn, p. 30.

IPIECA and UNEP. 1991. *Climate change and energy efficiency in industry.* London, IPIECA.

IULA (International Union of Local Authorities). 1991. *Glossary of environmental terms.* Istanbul, IULA-EMME.

Jähnicke, M., et al. 1989a. "Structural change and environmental impact", in *Intereconomics,* 24(1), Jan./Feb. 1989, pp. 24-35.

—. 1989b. "Economic structure and environmental impacts: East/West comparisons", in *The Environmentalist,* Sep. 1989, pp. 171-183.

James, P. 1996. "Quality and the environment: The experience of ICI", in North (ed.), 1996.

—. Bennett, M. 1994. *Environmental-related performance measurement in business.* Berkhamstead, Hertfordshire, Ashridge Management Research Group.

—. Stewart, S. 1994. *The European environmental executive – From technical specialist to strategic co-ordinator?* Berkhamstead, Hertfordshire, Ashridge Management Research Group.

Joint Industrial Safety Council (Sweden). 1987. *Safety, health and working conditions.* Training manual. Stockholm.

Keating, M. 1993, *Agenda for change.* Geneva, Centre for Our Common Future.

Kharbanda, O.P.; Stallworthy, E.A. 1990. *Waste management.* Aldershot, Hampshire, Gower.

Koos, F. 1991. *Eco-management, green accounting and environmental reporting* (mimeo). Geneva, ILO, Entrepreneurship and Management Development Branch.

Krol, A. 1995. "Environmental management – Issues and approaches for an organisation", in M.D. Rogers (ed.): *Business and the environment*. London, Macmillan, pp. 51-90.

Kubr, M. (ed.). 1996. *Management consulting*. Geneva, ILO, 3rd (revised) ed.

Leonard, H.J. 1988. *Pollution and the struggle for the world product. Multinational corporations, environment, and international competitive advantage*. Cambridge, Massachusetts, Cambridge University Press.

Newall, J.E. 1990. "Managing environmental responsibility", in *Business Quarterly*, Autumn 1990, pp. 90-94.

North, K. 1995. "Okologie beeinflusst immer mehr die Lieferantenwahl" [Ecology influences purchasing], in *Beschaffung Aktuell*, 11/95, pp. 63-70.

—. (ed.). 1996. *European casebook on environmental business management*. London, Prentice-Hall.

—. Daig, S. 1996. "Environmental training in UK and German companies", in W. Wehrmeyer (ed.): *Greening people*. Sheffield, South Yorkshire, Greenleaf Publishing.

Nulty, P. 1990. "Recycling becomes a big business", in *Fortune*, 13 Aug., pp. 41-45.

OECD (Organisation for Economic Co-operation and Development). 1991. *The state of the environment*. Paris.

Opschoor, J.B.; Vos, H. 1989. *The application of economic instruments for environmental protection in OECD member countries*. Paris, OECD.

Pachauri, R.K. 1990. "Energy efficiency and conservation in India", in *Industry and Environment*, 13 (1990), 2, pp. 19-24.

Pearce, D., et al. 1989. *Blueprint for a green economy*. London, Earthscan Publications.

Peters, T. 1991. "Lean, green and clean: The profitable company of the year 2000", in *EMD Environment*, 91/1, pp. 5-8.

Porter, M.E. 1990. "The competitive advantage of nations", in *Harvard Business Review*, 68, No. 2 (Mar./Apr.), pp. 73-93.

Porter, M.E.; van der Linde, C. 1995. "Green and competitive: Ending the stalemate", in *Harvard Business Review*, Sep.-Oct., pp. 120-134.

Prokopenko, J. 1987. *Productivity management*. Geneva, ILO.

Rada, J.F. 1990. *The Greening of the enterprise – Business leaders speak out on environmental issues*. Paris. International Chamber of Commerce (ICC).

RAL Deutsches Institut für Gütesicherung, 1990. *Umweltzeichen – The environmental label introduces itself*. Bonn.

REFA. 1985. *Methodenlehre des Arbeitstudiums*, Vol. 3. Munich, Hanser.

RIVM (National Institute of Public Health and Environmental Protection). 1989. *Concern for tomorrow*. Bilthoven.

Rossiter, C.; El Batawi, M.A. 1987. "The working environment", in *Industry and Environment*, 10 (1987), 3, pp. 3-11.

Royston, M. 1979. *Responsibility of industry towards the environment*, The Conoco Lecture. London, Conoco.

—. 1989. *Introduction to environmental management*, (draft manuscript). Geneva, ILO.

Sarnoff, 1971. *Encyclopedic dictionary of the environment*. New York, Quadrangle Books.

Schmidheiny, S., with the Business Council for Sustainable Development. 1992. *Changing course*. Cambridge, Massachusetts, MIT Press.

Sherman, S.P. 1989. "Trashing a $150 billion business", in *Fortune*, 28 Aug., pp. 64-68.

Simonis, U.E. 1989. "La modernisation écologique de la société industrielle: trois éléments stratégiques", in *Revue Internationale des Sciences Sociales* (121), Aug., pp. 383-399.

Taylor, B., et al. 1994. *The environmental management handbook*. London, Pitman Publishing.

ten Brink, P., et al. 1996. "Consulting the stakeholder: A new approach to environmental reporting for IBM (UK) Ltd.", in *Greener Management International*, No. 13, Jan., pp. 108-119.

The Economist. 1990. "Managing greenly". 8 Sep., p. 22.

—. 1990. "Seeing the green light". 20 Oct., pp. 88-89.

—. 1991. "Free trade's green hurdle". 15 June, pp. 69-70.

Tully, S. 1989. "What the greens mean for business", in *Fortune*, 23 Oct. 1989, pp. 46-52.

UNCED (United Nations Conference on Environment and Development). 1991. *Utilization of economic instruments*, Preparatory Committee for UNCED, Third Session, Progress report of the Secretary General of the Conference, Geneva.

UNEP (United Nations Environment Programme). 1988. *Environmental impact assessment: Basic procedures for developing countries*. Bangkok, UNEP Regional Office for Asia and the Pacific.

—. 1989a. *Register of international treaties and other agreements in the field of the environment* (regular update). Nairobi.

—. 1989b. *The state of the world environment* (published annually). Nairobi.

—. 1989c. *Environment Brief*, No. 7. Nairobi.

UNEP/IE (Industry and Environment). 1980. *Guidelines for assessing industrial environmental impact and environmental criteria for the siting of industry*. Paris.

—. 1988. *APELL*. Handbook. Paris.

—. 1990a. *Storage of hazardous materials: A technical guide for safe warehousing of hazardous materials*. Paris.

—. 1990b. *Environmental auditing*. Paris.

—. 1991. *Companies' organization and public communication on environmental issues*. Paris.

UNEP/IE. 1992. *From regulations to industry compliance: Building institutional capabilities*. Paris.

—. 1996. *Management of industrial accident prevention and preparedness: A training resources package*. Paris.

UNEP/IE/UNIDO (United Nations Industrial Development Organization). 1991. *Audit and reduction manual for industrial emissions and wastes*. Paris, UNEP/IE.

UNEP/WBCSD (World Business Council for Sustainable Development). 1996. *Eco-efficiency and cleaner production: Charting the course to sustainability*. Paris.

UNEP and Sustain Ability (Ltd.). 1994. *Company environmental reporting: A measure of the progress of business and industry towards sustainable development*. Paris, UNEP/IE/PAC.

United Nations Association in Canada. 1991. *Notes on the road to Brazil*. An information kit. Ottawa, Ontario.

Walley, N.; Whitehead, B. 1994. "It's not easy being green", in *Harvard Business Review*, May-June, pp. 46-52.

WCED (World Commission on Environment and Development). 1987. *Our common future*. Oxford, Oxford University Press.

Wehrmeyer, W. 1995. *Measuring environmental business performance – A comprehensive guide*. Cheltenham, Gloucestershire, Stanley Thornes.

—. 1996. "Hygiene factor or motivator? Environmental business management at Seaswift", in North (ed.), 1996.

Willums, J.O.; Golicke, U. 1992. *From ideas to action. Business and sustainable development*. Oslo, ICG Norway.

Winter, G. 1988. *Business and the environment*. Hamburg, McGraw-Hill.

Winter, G., et al. 1991. *Industry initiatives in achieving ecologically sustainable industrial development (ESID)*. International Conference on ESID, Copenhagen, Oct.

WHO (World Health Organization). 1982. *Rapid assessment of sources of air, water, and land pollution*. Offset Publication No. 62. Geneva.

—. 1983. *Management of hazardous waste*. WHO Regional Publications No. 14. Geneva, WHO Regional Office for Europe.

Womack, P., et al. 1990. *The machine that changed the world*. New York, Macmillan.

World Resources Institute. 1990. *World resources 1990-91*. Oxford, Oxford University Press.

2. Glossary

The environmental debate has developed its own jargon. This glossary is intended to assist managers in finding their way through the most important environmental terms without aiming at proving an exhaustive list.

Acid rain – Sulphuric acid particulates (aerosols) which are heavy enough to be washed out of the atmosphere in rain. Results from combustion of sulphur-rich fuels (e.g. coal-burning power plants) and considered to be harmful to plant and fish life. See Acidification.

Acidification – Acidification has been experienced mostly in Central Europe and Scandinavia. Forests are dying or are losing their vitality. Heathland grasses, fens and lakes are acidifying. High doses of sulphur and nitrogen compounds, especially in polluted cities, also have adverse health effects. Acidification is caused by emissions of sulphur and nitrogen compounds (SO_2, NO_x, NH_3) from large parts of the European continent. Emitters of SO_2 are power stations, refuse incinerators, refineries and private households. Road traffic mainly accounts for NO_x.

Administrative charges – Use of licences and fees to recover the costs of administering pollution control systems is widespread and applies to such things as the licensing of waste-disposal sites, the disposing of radioactive wastes and the control of chemicals.

Aerosol – Solid or liquid particulate matter suspended in the air because of its small size, e.g. smoke, smog, etc.

Air quality standards – The level of air pollution prescribed by law or regulation that cannot be exceeded during a specified time in a defined area. Ambient air quality standards refer to the maximum allowable levels of specific polluting materials permitted under the law.

Air, ambient – Surrounding outdoor air. That portion of the atmosphere, external to buildings, to which the general public has access.

Bhopal – An Indian town where, in 1984, the world's biggest chemical accident occurred. About 3,000 people were killed by toxic gas released from a Union Carbide plant and many more were disabled.

Biochemical Oxygen Demand (BOD) – Used as a measure of organic pollution and indicating the organic content of wastewater and surface water; also a measure of the effectiveness of sewage-treatment processes.

Biodegradation – The process of decomposing organic matter and substances as a result of the action of micro-organisms.

Biosphere – The portion of the earth and its atmosphere capable of supporting life. The thin covering of the planet that contains and sustains life.

BOD – See Biochemical Oxygen Demand.

Carbon taxes – To date, these taxes have operated generally as a combination of taxes on the carbon content of fuel (emission taxes) and taxes on energy consumption (product taxes).

Carcinogenic – Any chemical substance or form of energy (e.g. radiation) capable of producing cancer.

Chemicals – Chemical elements and compounds, and mixtures thereof, whether natural or synthetic, and whether in solid, liquid or gaseous/vapour form.

Chernobyl – A town in Ukraine, and site of the world's worst nuclear accident, in 1986.

Chloro-fluorocarbons (CFCs) – Highly stable compounds used in aerosol propellants, refrigeration, plastic foam and industrial solvents, believed to be a major factor leading to the depletion of the ozone layer and a contributor to the greenhouse effect.

Cleaner production – Production processes which cause low or no emissions and little or no waste, and which are energy efficient.

Depletion – Reduction, exhaustion, drying up of resources.

Deposit refund systems – Deposit refund systems operate like charges except that consumers pay a surcharge which is subsequently refunded when products (e.g. packaging materials) are returned to a collection system.

Desertification – Overgrazed, overstressed, very arid or fragile lands become deserts in many regions of the world. Rapid destruction of natural environments is reducing both the number of species and the amount of genetic variation within individual species. Biological diversity is therefore declining.

Dioxins – Extremely toxic substances found in herbicides; also formed when chlorinated organic materials such as PVC are burned at medium temperatures. See also Seveso.

Diversity of species – The biological complexity in numbers of species of organisms of an ecosystem. In many instances, the ecosystem becomes more stable as diversity increases.

Effluent and emission charges – The practice of applying charges to effluents and emissions, first introduced in a major way in France, to control water effluents has become widespread. There have also been some recent applications of charges on emissions of CFCs and nitrogen oxides.

Emission – A discharge of particulate, gaseous or soluble waste material/pollution into the air from a polluting source.

End-of-pipe treatment (abatement) – Instead of preventing the occurrence of polluting substances they are treated at the end of a process by, for example, filters, catalysts or scrubbers.

Environment – The sum of physical resources that sustain life and are a basis for satisfying human needs.

Environmental Audit (EA) – EA is a management tool to evaluate the environmental performance of a company, organization or parts thereof.

Environmental Business Management (EBM) – EBM means the integration of environmental protection into all managerial functions with the aim of reaching an optimum between the economic and ecological performance of a company.

Environmental charges and taxes – The main objective of this instrument is to recover some of the costs of pollution control consistent with the polluter pays and user pays principles, while generating revenue to finance the growing cost of environmental protection.

Environment-friendly – Term used for products which at all stages of their life cycle are not, or only minimally, harmful to the environment, often also called Green products.

Environmental awareness – The growth and development of awareness, understanding, and consciousness toward the biophysical environment and its problems, including human interactions and effects. Thinking "ecologically" or in terms of an ecological consciousness.

Environmental degradation – Any action which makes the environment less fit for human, plant, or animal life. Also associated with the lowering and reduction of environmental quality.

Environmental impact assessment (EIA) – The EIA is a tool for decision-makers to forecast the impact that a project will have on the environment and to find ways to reduce unacceptable impacts.

Environmental management – Use and protection of natural resources through the application of environmentally sound practices, including the management of land, water and air.

Environmental management system (EMS) – An EMS consists of an environmental policy, targets, objectives and organizational provisions for implementing environmental measures, and dispositions for monitoring and reporting. The requirements for an EMS are stipulated by, for example, BS 7750, the ISO 14000 Series or the European Environmental Management Audit Scheme (EMAS).

Eutrophication – In many countries soil, groundwater, lakes, rivers and coastal zones are heavily polluted by the nutrients nitrogen, phosphorus and potassium. This pollution is caused by agricultural fertilizers and deposits of nitrogen from the air. In coastal areas, lakes and rivers, these excessive nutrient loads lead to higher densities of vegetable organisms (algae boom). This can destabilize the aquatic ecosystem, resulting in a decline of aquatic species. Another alarming result is that the nitrate concentration in ground water is rising, which constitutes a threat to the drinking-water supply.

Exposure limit – Exposure limits, the definition and legal status of which vary from country to country, refer to concentrations or intensities at the workplace which in repeated long-term exposure, even up to an entire working life, do not in general lead to health impairment of either the workers or their offspring.

Genetic diversity – The genetic materials associated with a variety and number of species of organisms. Protection of genetic diversity is essential to sustain and improve agriculture, forestry, and fisheries, to keep open future options, and to provide for a buffer against harmful change.

Global commons – Land (Antarctica), water, or air owned or used jointly by the members of the community of nations. Global commons include those parts of the earth's surface beyond national jurisdictions, including the atmosphere.

Global warming – See Greenhouse effect.

Greenhouse effect – The greenhouse effect is brought about globally by emissions of CO_2 (energy, deforestation), methane (agriculture), nitrous oxide (energy), CFCs (chloro-fluorocarbons from industrial products) and by ozone. These gases lead in the troposphere (at an altitude of 2-10 km) to a change in the radiation of heat from the atmosphere into space. As a result, it is forecasted that the atmosphere will warm, the sea level will rise and climates will change worldwide.

Habitat – The sum of environmental conditions in a specific place that is occupied by an organism, population or community, and where it naturally lives and grows.

Hazardous chemicals – Include any chemical which has been classified as hazardous in accordance with Article 6 of the ILO Convention concerning Safety in the Use of Chemicals at Work (No. 170) or for which relevant information exists to indicate that the chemical forms a threat to the environment or to human health.

Hazardous wastes – Hazardous wastes have physical, chemical and biological characteristics which require special handling and disposal procedures to avoid risk to health and/or other adverse environmental effects. The composition of hazardous waste varies considerably from one industry to another, and so the composition of waste generated in different countries varies with different groupings of industry.

Heavy metals – A group of metals, some of which are essential for healthy life (zinc, copper, manganese) while others (cadmium, lead) are toxic.

Hydrocarbons – Organic compounds consisting only of carbon and hydrogen which are found commonly in fossil fuels and in the products of partial combustion of these substances, such as in the exhaust gases of gasoline-driven vehicles. Methane (CH_4) contributes to the greenhouse effect.

Incineration – Process of controlled burning or oxidation of combustible waste, thus reducing waste volume by 80 to 90 per cent. See Dioxins.

Montreal Protocol – The Montreal Protocol is concerned with phasing out ozone layer depleting gases:

> For methyl chloroform 70 per cent reduction by 2000
>> 100 per cent reduction by 2005.
>
> For carbon tetrachloride 85 per cent reduction by 1995
>> 100 per cent reduction by 2000.
>
> For halons 50 per cent reduction by 1995
>> 100 per cent reduction by 2000
>> (exceptions for a few special users).

NIMBY – Not In My Backyard. Description of phenomenon that people would agree to a project as long as it is not near to their homes.

Nitrogen oxides (NO_x) – These are three main oxides of nitrogen (NO, NO_2 and N_2O_5) produced in varying amounts by natural processes (nitrification and decomposition) and during the combustion of fossil fuels, for example in furnaces and motor vehicles. The oxides are generally not differentiated, but grouped together as NO_x, which contribute to atmospheric pollution such as acid rain and smog, and are therefore subject to emission controls.

Non-compliance fees – These are imposed if polluters fail to comply with required standards. In Mexico, for example, such fees are adjusted to inflation in order to remove any possible financial reason for seeking to delay payment.

Non-waste technology – Technology which allows production with no or little waste resulting (e.g. by total recycling of by-products, closed water and air circuits). See also Cleaner production.

Nuclear radiation – The nuclear accidents at Three Mile Island (United States) and Chernobyl made clear that the present state of nuclear power generation is a major threat to the environment and human life. Not only will nuclear accidents lead to environmental pollution but there is always the problem of radioactive wastes and regular discharges. The areas in which humanity is exposed to the released substances vary greatly in terms of the dispersion. The greatest exposure takes place via:

– External radiation by gamma rays from radio nuclides in the radiation cloud and on the ground.

– Interior radiation after inhalation of radioactive aerosols and after ingestion of contaminated food, milk and water.

Due to the Chernobyl accident there have been many acute victims and more will die of cancer ten years or more after the accident.

Ozone at ground level – Ozone (O_3) is a very reactive gas and it is considered to be one of the most powerful oxidizing agents. This causes ozone to react quickly in the soil and water and explains why it does not play a significant role there. Primary emission of ozone, of a natural or anthropogenic nature, hardly ever occurs. Ozone is a secondary air pollution component that is formed in the atmosphere. Various processes are involved in the production of ozone, depending on altitude. In terms of the risk to mankind, the effects on the respiratory tract (diminished lung function, increased sensitivity to respiratory infections) are estimated to be the most serious.

Ozone depletion – Chloro-fluorocarbons (CFCs) and halons damage the ozone layer in the stratosphere. The ozone layer absorbs UV-B radiation which is harmful to humans and nature. Increasing UV-B radiation can result in serious damage to public health from skin cancer and a weakening of the immune system, so that infectious diseases may develop. Satellite observations show that the decrease in ozone becomes greater every year. The loss of ozone lessens with increasing distance from the poles. CFCs are used as a refrigerant in refrigerators and air-conditioning units, a propellant in aerosol sprays and a blowing agent in the production of synthetic materials. Halons are used in fire extinguishers. The Montreal Protocol foresees the phasing out of most CFCs by the year 2000.

Particulate matter – Visible or invisible, solid or liquid, finely divided parts that can be suspended in gas or air, such as dust, sand or ashes.

Performance bonds – To enforce compliance with regulatory requirements, performance bonds are imposed before activity begins, as for example in the case of offshore drilling in Canada and the United States.

Persistent substances – Persistent substances, such as metals, pesticides and anorganic compounds, likewise constitute a major threat to the quality of soil and groundwater lakes, rivers and seawater, which are increasingly loaded with metals from the use of fertilizers and industrial production, and deposition from the air. The cadmium and copper concentrations in the soil are rising. The consumption of pesticides in agriculture has grown. As a result the concentrations of pesticides are increasing. Mobile pesticides leak into groundwaters. They pose a threat to the drinking-water supply.

Pesticide – Any chemical substance used to kill plant and animal (and insect) pests. Some pesticides can contaminate water, air, or soil and can accumulate in humans, plants, animals and the environment with negative effects.

Pet (polyethylene trephtalate) – A recyclable plastic used mainly in mineral water and other soft drinks bottles.

Pollutant – Any extraneous material or form of energy whose rate of transfer between two components/factors of the environment is changed so that the well-being of organisms/ecosystems is negatively affected. Any introduced gas, liquid, or solid that makes a resource unfit for a specific purpose or that adversely affects human, plant or animal life.

Pollution – The presence of matter or energy whose nature, location or quantity produces undesirable environmental effects. The contamination or alteration of the quality of some portion or aspect of the environment and its living organisms by the addition of harmful impurities.

Pollution control – Systems of measurement, criteria, standards, laws, and regulations which are directed at the sources and causes of various forms of pollution and its effects in terms of control and prevention. Control measures involve both quantity (degree) and quality (value) considerations.

Product charges – Charges are being applied to an increasing range of environmentally damaging products. These can be applied to inputs in the production process, such as the charges which have been applied by some European countries to fertilizers and pesticides. Similarly, they can be attached to products or services sold, such as the growing number of charges which are being applied to items which pose particularly difficult or costly disposal problems, such as waste oil, tyres, lead acid batteries and non-returnable containers.

Product life-cycle analysis – An evaluation of a product's environmental impact considering all stages of its life cycle from cradle to grave, also called ecological (eco) balance.

PVC (polyvinyl chlorate) – A plastic used to make, for example, pipes, mineral water bottles, floor coverings and packaging materials. The use of PVC is now becoming widely restricted due to the carcinogenic character of the raw materials from which it is made.

Recycling – Process of reusing discarded or waste materials. Up to now, mainly materials such as paper, plastics, glass and chemical solvents are recycled. An efficient separation of wastes is a condition for recycling.

Renewable energy – Energy sources not diminished by use, such as solar energy, wind power, geothermal power, hydropower, biomass energy, tidal power and breeder nuclear reactors.

Renewable resources – Resources that, if properly managed, will naturally replace themselves or may be regenerated through human interventions, such as crops, cattle, trees, air or water.

Scrubber – Device to counter air pollution consisting of an atomized spray of water that collides with undesirable particles, traps them and separates them from the emission which then flows unpolluted up to the stack.

Seveso – Italian town where a major chemical explosion occurred in 1976. Dioxin was released and the subsequent explosion led to the so-called Seveso Directive of the European Communities stipulating that certain factories provide the emergency services with details of all chemicals stored on their sites.

Sludge – Concentration of solids resulting from (sewage) treatment, produced by filtration, sedimentation and/or biological treatment.

Smog – Photochemical haze mostly in urban areas caused by automobile exhausts and other emissions undergoing photochemical reaction in the atmosphere. Smog can cause eye irritation, respiratory ailments, plant damage and reduced visibility. See also Ozone at ground level.

Soil erosion – Soil erosion has been reported from almost every country in the world. Declines in soil fertility – or even total losses of land to agriculture – are common in many parts of the world. Salinization, for example, affects extensive land areas in many countries in North Africa, the Middle East and Asia. About half the land under irrigation is affected by secondary salinization and/or alkalization in varying degrees.

Solid waste – Also known as refuse, any non-liquid discarded material.

Solvents, industrial – The expression "industrial solvents" is conventionally applied to organic liquids capable of dissolving a large number of substances. Solvents may be hydrocarbons, alcohols, ethers, glycol derivates, etc. Solvents are used in industry for a variety of purposes: different forms of surface coatings are made possible by solvents, such as printing inks and paints and spreading preparations for paper and fabrics of various types. Solvents are used to reduce materials to a plastic condition in order that they may be moulded, extruded or otherwise shaped. They are used for extracting oils, fats and medicinal material from seeds, nuts and bones, etc. Fire, explosion and health hazards must be considered when industrial solvents are manufactured and used. Further, the property of causing narcosis is common to most organic industrial solvents.

Sulphur dioxide (SO_2) – A colourless, irritating pungent gas formed when sulphur burns in air, one of the major air pollutants which contributes to acid rain and smog. Originates in the combustion of the sulphur in most fossil fuels.

Superfund – As part of the Comprehensive Environmental Response, Compensation and Liability Act in the United States a superfund was constituted to pay for the clean-up of contaminated industrial areas.

Sustainable development – Sustainable development is development that meets the needs of the present without compromising the ability of future generations to meet their own needs.

Thermal pollution – The introduction of heated effluent into water, raising the temperature to the point where it is detrimental for the aquatic environment, for example, by increasing proliferation of algae.

Threshold limit value (TLV) – See Exposure limit.

Tradable pollution permits – Also called marketable permits or emission trading. The development of tradable permit systems in principle involves setting limits on total allowable emissions within a geographic area; permits are then allocated to industry which is permitted to trade them with other companies or to offset higher emissions in their own facilities. Assuming that firms minimize their total production costs and that the market for these permits is competitive, the resulting level of pollution control will be achieved in the most cost-effective manner.

Tropical Timber Agreement – The Tropical Timber Agreement which came into force in 1985 under the auspices of UNCTAD is now implemented by the International Tropical Timber Organization (ITTO) established at Yokohama in Japan in 1987. ITTO's main objectives are to improve market intelligence, to assist producing countries to develop better techniques for reforestation and forest management, to encourage increased timber processing in producing countries, and to support research and development programmes to achieve these goals.

User charges – There is a growing practice of using charges to recover not only administrative costs but also the costs of pollution control. Thus far such charges have been applied mainly to sewage and water management. There is growing interest in making greater use of solid waste pricing systems to provide incentives for waste reduction, although care needs to be taken to ensure these are not set so high that they encourage illegal dumping.

Valdez principles – Principles of environmental management developed by investment groups in the United States following the *Exxon Valdez* oil spill in Alaska. These principles must be met by a public company before the investors buy shares in a company.

Waste – Any substance or object which the holder disposes of or is required to dispose of pursuant to the provisions of national law in force.

Waste, biodegradable – Organic waste materials (industrial and domestic) that can be broken down (decomposed) into their basic elements by the action of micro-organisms.

Waste, non-biodegradable – Inorganic or mineral waste materials (industrial and domestic) that cannot be broken down (decomposed) into their basic elements by the action of micro-organisms and, consequently, remain in the environment for a long or indefinite period of time, e.g. synthetic materials (plastics), metal, etc.

Waste tourism – A term used for transporting wastes within countries or across borders due to less strict environmental rules in the waste-receiving region or country. On 29 January 1991, 12 African nations adopted a treaty, the Bamako Convention, which will close their countries to all forms of hazardous waste trade.

Water quality standards – Characteristics and degrees/levels of water quality for comparison in terms of different sources and uses. A management plan that considers: (a) water uses; (b) setting water quality criteria levels to protect those uses; (c) implementing and enforcing the water treatment plans; (d) protecting existing high-quality waters; and (e) establishing regulations designating standards for bodies of water.

Working environment – The working environment comprises physical environmental factors such as noise, climate, illumination, vibrations, air quality as well as work organization.

Zero pollution – The objective to reduce emission to a level (near) zero. Often used as a slogan of company programmes for pollution prevention.

Sources of glossary

1. Henning (No date).
2. IULA, 1991.
3. Sarnoff, 1971.
4. RIVM, 1989.
5. Elkington et al., 1991.
6. ILO.
7. UNEP, 1989b.
8. WCED, 1987.
9. UNCED, 1991.
10. Brackley, 1988.

3. Useful addresses

Employers' and business organizations

European Roundtable of Industrialists
Secretariat Office
Rue Guimard 15
B-1040 Brussels
Belgium
Tel.: 322/5115800

GEMI (Global Environmental Management Initiative)
1828 L Street, NW Suite 711
Washington, DC 20036
United States
Tel.: 202/296-7449

International Chamber of Commerce
38, Cours Albert 1er
75008 Paris 10900
France
Tel.: 331/45 62 34 56

International Labour Office (ILO)
Employers' Activities (ACT/EMP)
CH-1211 Geneva 22
Switzerland
Tel.: 022/799 77 17

The International Network for Environmental Management
c/o BAUM
Christian-Förster-Str. 19
D-2000 Hamburg 13
Germany
Tel.: 040/40 77 21

International Organization of Employers
28, chemin de Joinville
CH-1216 Cointrin
Switzerland
Tel.: 022-698 16 16

World Business Council for Sustainable Development (WBCSD)
World Trade Centre
3rd, floor
10, route de l'Aéroport
CH-1215 Geneva 15
Switzerland
Tel.: 022/788 32 02

Workers' organizations

International Confederation of Free Trade Unions
37-41, rue Montagne aux Herbes Potagères
B-1000 Brussels
Belgium

International Federation of Building and Woodworkers (IFBWW)
20, rte de Pré-Bois
PO Box 733
CH-1215 Geneva 15, Aéroport
Switzerland
Tel.: 022/788 08 88

International Federation of Chemical, Energy, Mine and General Workers' Unions
109, avenue Emile de Béco
B-1050 Brussels
Belgium
Tel.: 322/647 02 35

International Labour Office (ILO)
Bureau for Workers' Activities (ACTRAV)
CH-1211 Geneva 22
Switzerland
Tel.: 022/799 70 21

International Metalworkers' Federation (IMF)
54, bis route des Acacias
PO Box 563
CH-1227 Carouge
Switzerland
Tel.: 022/43 61 50

Miners' International Federation (MIF)
109, avenue Emile de Béco
B-1000 Brussels
Belgium
Tel.: 00322 646 21 20

World Confederation of Labour
33, rue de Trèves
B-1040 Brussels
Belgium
Tel.: 00322 230 60 90

World Federation of Trade Unions
35, Na Dobesce
140 00 Prague 4
Czech Republic
Tel.: 422 46 28 84

Environmental groups

African NGOs Environment Network
PO Box 53844
Nairobi
Kenya
Tel.: 2542/26255

Bund für Umwelt-und Naturschutz
Postfach 120536
Bonn D-5300
Germany

Earthscan
3, Endsleigh Street
London WC1H 0DD
United Kingdom
Tel.: 171/388 2117

Environment and Development in the Third World
BP 3370
Dakar
Senegal
Tel.: 221/224229

Environment Liaison Centre International
PO Box 72461
Nairobi
Kenya
Tel.: 2542/24770

European Environmental Bureau
20, rue du Luxembourg
B-1040 Brussels
Belgium
Tel.: 322/5141432

Friends of the Earth
218, D St., SE
Washington, DC 20003
United States
Tel.. 202/5442600

Greenpeace
35, Graham Street
London N1 8ll
United Kingdom
Tel.: 171/251 3020

International Union for Conservation of Nature and Natural Resources (IUCN)
Avenue du Mont-Blanc
CH-1196 Gland
Switzerland
Tel.: 022/647181

National Wildlife Federation
1400, 16th St., NW
Washington, DC
United States
Tel.: 202/797 6800

World Conservation Union (IUCN)
Avenue du Mont-Blanc
CH-1196 Gland
Switzerland
Tel.: 022/647181

World Resources Institute
1709, New York Ave., NW
Washington, DC 20006
United States
Tel.: 202/638 6300

Worldwatch Institute
1776, Massachusetts Ave., NW
Washington, DC 20036
United States
Tel.: 202/452 1999

World Wide Fund for Nature (WWF)
Avenue du Mont Blanc
CH-1196 Gland
Switzerland
Tel.: 22/64 91 11

International organizations and institutions

African Ministerial Conference of the Environment
Bureau of the Conference (UNITERRA)
c/o UNEP
Regional Office/Africa
PO Box 30552
Nairobi
Kenya
Tel.: 2542/33930

Caribbean Environmental Health Institute (CEHI)
PO Box 1111
The Morne Castries
St. Lucia
Tel.: 1809/4522501

Centre for our Common Future
Palais Wilson
52, rue des Pâquis
CH-1201 Geneva
Switzerland

Club of Rome
c/o IBI
PO Box 10253
00144 Rome
Italy

Committee of International Development Institutions on the Environment (CIDIE)
PO Box 30552
Nairobi
Kenya
Tel.: 2542/333930

Food and Agriculture Organization of the United Nations (FAO)
Environment and Sustainable Development Programmes
Coordinating Centre
Via della Terme di Caracalla
000100 Rome
Italy
Tel.: 39/6/57971

International Academy of Environment
4, chemin de Conches
CH-1231 Conches/Geneva
Switzerland
Tel.: 022/789 13 11

International Labour Office (ILO)
Entrepreneurship and Management Development Branch (ENT/MAN)
CH-1211 Geneva 22
Switzerland
Tel.: 022/799 68 56

International Occupational Safety and Health Information Centre (CIS)
International Labour Office
CH-1211 Geneva 22
Switzerland
Tel.: 022/799 67 39

International Organization for Standardization (ISO)
Case Postale 56
CH-1211 Geneva 20
Switzerland
Tel.: 22/7341240

Pan American Centre for Sanitary Engineering and Environmental Sciences
Casilla Postal 4337
Lima 100
Peru
Tel.: 5114/354135

Regional Network of Non-Governmental Conservation Organisations for Sustainable Development in Central America
Apartado Postal 2431
Managua
Nicaragua
Tel.: 5052/3765-4448

South Asia Cooperative Environmental Programme (SACEP)
PO Box 1070
Colombo 5
Sri Lanka

United Nations Development Programme (UNDP)
Environment and Natural Resources Unit
One United Nations Plaza
New York, New York 10017
United States
Tel.: 212/906.5000

United Nations Environment Programme (UNEP)
Industry and Environment
Tour Mirabeau
39-43, quai André Citroën
75730 Paris
France
Tel.: 331/40 58 88 56

UNEP IE
APELL Programme Coordinator
39-43 quai André Citroën
75739 Paris
Fax.: 331/44 37 14 74
E-mail: unepie@unep.fr

UNESCO
9, place de Fontenoy
75007 Paris
France
Tel.: 331/45 68 10 00

United Nations Industrial Development Organization (UNIDO)
Environment Unit
Vienna International Centre
PO Box 400
A-1400 Vienna
Austria
Tel.: 211/31.3956

World Bank
Environment Department
1818, H Street NW
Washington, DC, 20433
United States
Tel.: 202/477.1234

World Health Organization
20, avenue Appia
CH-1211 Geneva 27
Switzerland
Tel.: 022/791 21 11

4. International Chamber of Commerce Business Charter for Sustainable Development Principles for Environmental Management, adopted by the ICC Executive Board on 27 November 1990

Foreword

There is widespread recognition today that environmental protection must be among the highest priorities of every business.

In its milestone 1987 report, "Our common future", the World Commission on Environment and Development (Brundtland Commission) emphasized the importance of environmental protection in the pursuit of sustainable development.

To help business around the world improve its environmental performance, the International Chamber of Commerce established a task force of business representatives to create this Business Charter for Sustainable Development. It comprises 16 principles for environmental management which, for business, is a vitally important aspect of sustainable development.

This Charter will assist enterprises in fulfilling their commitment to environmental stewardship in a comprehensive fashion. It was formally launched in April 1991 at the Second World Industry Conference on Environmental Management.

Introduction

Sustainable development involves meeting the needs of the present without compromising the ability of future generations to meet their own needs. Economic growth provides the conditions in which protection of the environment can best be achieved, and environmental protection, in balance with other human goals, is necessary to achieve growth that is sustainable.

In turn, versatile, dynamic, responsive and profitable businesses are required as the driving force for sustainable economic development and for providing managerial, technical and financial resources to contribute to the resolution of environmental challenges. Market economies, characterized by entrepreneurial initiatives, are essential to achieving this.

Business thus shares the view that there should be a common goal, not a conflict, between economic development and environmental protection, both now and for future generations. Making market forces work in this way to protect and improve the quality of the environment – with the help of performance-based standards and judicious use of economic instruments in harmonious regulatory framework – is one of the greatest challenges that the world faces in the next decade.

The 1987 report of the World Commission on Environment and Development, "Our common future", expresses the same challenge and calls on the cooperation of business in tackling it. To this end, business leaders have launched actions in their individual enterprises as well as through sectorial and cross-sectorial associations.

In order that more businesses join this effort and that their environmental performance continues to improve, the International Chamber of Commerce hereby calls upon enterprises and their associations to use the following principles as a basis for pursuing such improvement and to express publicly their support for them. Individual programmes developed to implement these principles will reflect the wide diversity among enterprises in size and function.

The objective is that the wide range of enterprises commit themselves to improving their environmental performance in accordance with these principles, to having in place management practices to effect such improvement, to measuring their progress, and to reporting this progress as appropriate internally and externally.

Note: The term "environment" as used in this document also refers to environmentally related aspects of health, safety and product stewardship.

Principles

1. **Corporate priority:** To recognize environmental management as among the highest corporate priorities and as a key determinant to sustainable development; to establish policies, programmes and practices for conducting operations in an environmentally sound manner.

2. **Integrated management:** To integrate these policies, programmes and practices fully into each business as an essential element of management in all its functions.

3. **Process of improvement:** To continue to improve corporate policies, programmes and environmental performance, taking into account technical developments, scientific understanding, consumer needs and community expectations, with legal regulations as a starting-point; and to apply the same environmental criteria internationally.

4. **Employee education:** To educate, train and motivate employees to conduct their activities in an environmentally responsible manner.

5. **Prior assessment:** To assess environmental impacts before starting a new activity or project and before decommissioning a facility or leaving a site.

6. **Products and services:** To develop and provide products or services that have no undue environmental impact and are safe in their intended use, that are efficient in their consumption of energy and natural resources, and that can be recycled, reused, or disposed of safely.

7. **Consumer advice:** To advise, and where relevant educate, customers, distributors and the public in the safe use, transportation, storage and disposal of products provided; and to apply similar considerations to the provision of services.

8. **Facilities and operations:** To develop, design and operate facilities and conduct activities taking into consideration the efficient use of energy and materials, the

sustainable use of renewable resources, the minimization of adverse environmental impact and waste generation, and the safe and responsible disposal of residual wastes.

9. **Research:** To conduct or support research on the environmental impacts of raw materials, products, processes, emissions and wastes associated with the enterprise and on the means of minimizing such adverse impacts.

10. **Precautionary approach:** To modify the manufacture, marketing or use of products or services or the conduct of activities, consistent with scientific and technical understanding, to prevent serious or irreversible environmental degradation.

11. **Contractors and suppliers:** To promote the adoption of these principles by contractors acting on behalf of the enterprise, encouraging and, where appropriate, requiring improvements in their practices to make them consistent with those of the enterprise; and to encourage the wider adoption of these principles by suppliers.

12. **Emergency preparedness:** To develop and maintain, where significant hazards exist, emergency preparedness plans in conjunction with the emergency services, relevant authorities and the local community, recognizing potential transboundary impacts.

13. **Transfer of technology:** To contribute to the transfer of environmentally sound technology and management methods throughout the industrial and public sectors.

14. **Contributing to the common effort:** To contribute to the development of public policy and to business, governmental and intergovernmental programmes and educational initiatives that will enhance environmental awareness and protection.

15. **Openness to concerns:** To foster openness and dialogue with employees and the public, anticipating and responding to their concerns about the potential hazards and impacts of operations, products, wastes or services, including those of transboundary or global significance.

16. **Compliance and reporting:** To measure environmental performance; to conduct regular environmental audits and assessments of compliance with company requirements, legal requirements and these principles; and periodically to provide appropriate information•to the Board of Directors, shareholders, employees, the authorities and the public.

Index*

*Page numbers in *italics* refer to figures, in **bold** to tables.